WHAT IS EVOLUTION?
UNDEFINED. UNSCIENTIFIC. UNLAWFUL.

WILLIAM JAMES HERATH

WHAT IS EVOLUTION?
UNDEFINED. UNSCIENTIFIC. UNLAWFUL.

Copyright © William James Herath, 2017
All rights reserved.

This book is protected by the copyright laws of the United States of America, and may not be copied or reprinted for commercial gain or profit.

Design by William James Herath

Cover photo by unknown photographer.
Cover photo colorized by Mads Madsen.
Back cover photo by Scott Cerny.

www.ReadySetQuestioncom

ISBN-13: 978-1511659635
ISBN-10: 1511659637

87654321-0-12345678 // 87654321-0-12345678

DEDICATED TO THE FUTURE OF:

Scientific Progress, Innovation, Education & Civil Liberties

CONTENTS

Introduction   *i*

chapter 1. **WHAT IS EVOLUTION?**   1
    Creating a Definition
    Result and Means

chapter 2. **WHAT ARE SPORES OF CONFUSION?**   15
    Domestication
    Natural Selection
    Adaptation
    Trait Variation
    Speciation

chapter 3. **IS EVIDENCE EVIDENT?**   25
    Diversity of Species
    The Fossil Record
    Punctuated Equilibrium
    Genetic Island Formation
    Living Fossils
    Homology
    The Mythical Neanderthal
    Mutation
    Genetic Variation
    Vestigial Remnants
    Increased Genetic Complexity
    Junk DNA
    Our Closest Relative

chapter 4. **WHAT EMPIRICAL DATA SUPPORTS EVOLUTION?**   77
    Defining Science
    Scientific Prediction
    Observability of Evolution
    Processes & Mechanisms
    Falsifiability
    Inference
    Asking for a Hoax
    Questioning Evolution

chapter 5. **IS FAITH NECESSARY?**   103
    Healthy Competition
    The Magic of Millions
    Religions that Coexist

chapter 6. IS SPENDING TAX DOLLARS ON
            EVOLUTION LEGAL?   111
                    Legal Precedent
                    Evolution is Supernatural
                    Eradication of Opposition

Conclusion   *123*

References   *131*

Thank You   *141*

INTRODUCTION

"Live as if you were to die tomorrow. Learn as if you were to live forever."
Mahatma Gandhi

I was the annoying kid with way too many questions. Let's be honest, the world we live is fantastic and as a kid I really needed to ask questions. There was an inquisitive burn deep down inside and I am sure it was tiring for those around me. My folks had to have been exhausted from the barrage of questions, but they were so kind by describing my behavior as having a "real thirst for knowledge."

Growing up in a pre-Google age, information was acquired by the most archaic of means. My only options were to find an expert, go to the library, or be fortunate enough to randomly discover a Nova special that happened to address the question I had. So, in short, I exhausted my only real option of constantly demanding my parents' mental energy. Then one glorious day, my inquisitive dreams came true. Mom and dad had purchased the entire Encyclopedia Britannica! I still remember the warm scent of finished

WHAT IS EVOLUTION?

hide spreading through our home. Each burgundy, leather bound book embossed with golden letters seemed to woo questions from my soul. There was a weight to the books which gave such a feel of quality. A cracking and creaking sound would escaped as I joyfully opened the cover. Gently thumbing through the thin, translucent pages invoked a real sense of reverence. The experience of attaining knowledge by means of an actual Encyclopedia Britannica caused me to value each word just a little more.

After that glorious day, mom and dad responded to my incessant need for information by a liberating response of, "Let's go look it up!" Soon, I learned to skip the middleman and do research on my own. Of which, more than likely, caused my folks to belch out a guttural cry of freedom similar to Mel Gibson's portrayal of William Wallace in Braveheart.

In 2004, I became a dad. In 2005, I became a dad a second time over. In 2006, I began working with high school and middle school students, which is like being a dad a fifty times over. When you're a dad, you get asked a lot of questions. My kids are just like I was and so are my students. Payback is real!

Over the years, I have been asked thousands of questions.

INTRODUCTION

Relational stuff usually dominates the FAQ's. Most youth need guidance in regard to struggles they are having with a peer, a boyfriend, a girlfriend, or their parents. Yet, another frequently asked question has to do with the "how" of "How we got here?" Of course the "we" is referring to us, you know, we humans. Believe it or not, some kids have moments of deep reflection. In a world with instant entertainment at their fingertips, surprisingly, many teens and tweens brood over their own existence. I have personally been asked to clarify the intricacies of evolution dozens of times.

In public school, students are taught evolutionary ideas as unquestionable truth, which puts a major cavity of sorts on their faith in a Creator. Research polls have repeatedly shown evolution to be a major factor in determining why young people walk away from their faith, and as a youth pastor I had to respond. Unfortunately, I had no idea how to respond, so in order to better serve my students I dove in head first. I wanted to know everything there was to know about evolution. The question of "What is evolution?," rocketed me into a new season of life. Some might assume that a youth pastor would have picked up Creationist materials, but I wanted to know what evolution was. So, I began to read scientific journals like Nature, Science, Evolution, The Proceedings of the National Academy of

WHAT IS EVOLUTION?

Sciences, and The Journal of the American Medical Association. I set up Google alerts for the latest articles and findings in the evolutionary community. I read Darwin's Origin of Species over and over. I watched documentaries and read the New York Time's best books on evolution. I went as far as to contact heads of Biology Departments at major universities. I was getting an education in biological evolution and I began to feel a bit like Alice tumbling into the rabbit hole.

While flooding my mind with evolutionary ideas, I discovered a curious inconsistency. No journal, publication, book, documentary, or professor offered a scientifically agreed upon definition of evolution. In fact, some descriptions I found were in oblate contrast to each other. Yes, even professors at prestigious universities (of which I will leave nameless) have sent me their favorite or personal definitions of evolution which conflict and/or minimally coincide with other professors' ideas of evolution. Throughout all of my research I have yet to find a standard, scientifically agreed upon definition of evolution.

How could this be? Evolution is a compulsory subject in U.S. public schools, so the educational system must have an agreed upon definition, right? Well, I contacted the federally funded Next Generation Science Standards which

INTRODUCTION

has created our national science education framework and I was shocked to discover it does not offer a definition. Then I contacted the California Department of Education and received the same response. Just like the Next Generation Science Standards, the California Department of Education requires the teaching of evolution, but does not offer a definition. Next, I contacted the Los Angeles Unified School District, which is the second largest school district in the United States after the New York City Unified School District. Again, I was shocked to discover the Los Angeles Unified School District also requires evolution to be taught, but does not offer a definition.

Being completely baffled, I turned to high profile court cases in hopes of finding a definition of evolution. We all know there has been legal contention surrounding the evolutionary debate for years, there must be a legal definition, right? I began reading every court transcript dealing with evolution going all the way back to the 1925 Scopes trial. I was sure there would be a definition of evolution. Yet, the mystery of the rabbit hole got a bit deeper and darker. To my knowledge, every U.S. court case that has dealt with evolution have all taken place in the absence of a consistent, legal, and scientifically agreed upon definition.

WHAT IS EVOLUTION?

How could such a contentiously argued concept have become part of our nation's required science curriculum, yet been left without a definition? Is biological evolution intentionally undefined? If there is a reason or a purpose as to why evolution does not have a definition, what is it? Is the evolutionary community trying to hide something? I am starting to feel like an annoying little kid again. The current state of evolutionary ideas and education cause me to ask so many questions, but this time the Encyclopedia Britannica is not going to help.

Thank you for spending the time to educate yourself on this subject. I have taken much care and effort in writing this book and all information covered has been meticulously referenced. Do yourself a favor and read this book in its entirety. Also, look up the data and cross check the references, because you need to see the truth for yourself.

You are about to embark upon a journey of discovery. There will be moments of frustration, but discomfort should be expected during a shift in paradigm. I am glad you are joining me in asking questions. Welcome to the shocking reality that evolution is undefined, unscientific, and unlawful.

Chapter 1.

WHAT IS EVOLUTION?

"Evolution is almost universally accepted among those who understand it, almost universally rejected by those who don't."
Richard Dawkins

When it comes to how we feel about evolution, these United States are divided. According to a 2014 survey, the Pew Research Center discovered that 33% of Americans say humans "evolved; due to natural processes," 25% say humans "evolved; due to God's design," 34% of Americans say that humans have "always existed in present form," 4% say humans "evolved; but don't know how," and another 4% just say "they don't know."

It is impossible for all opinionated groups to be correct and almost equally impossible for all groups to agree. Countless arguments have been heard on numerous courtroom floors. It is clear to see that generation after generation, the heated controversy around evolution has continued and is continuing headstrong into the foreseeable future. The public classroom and what is taught to the next generation has proven to be the prize most who enter the battle seek

WHAT IS EVOLUTION?

to win. Celebrities have been made over this debate, yet taxpayers have lost out by flipping the tab for a train of perpetual trials.

When it comes to evolution, does anyone in the United States have a clear grasp of what they are standing up for or standing up against? Evolution is a word that most presuppose comprehension. When asked, an opinion is quickly shared and regrettably all parties involved tend to befuddle each other. If one were to claim our populace to have a good working definition of evolution and those who support or oppose its validity communicate from that same interpretation, they would be speaking erroneously.

In using the word with regard to explaining biodiversity, a myriad of uses pertaining to various ideas and natural occurrences are enveloped. Webster is the go to source that many have used when in need of clarity and defines evolution to be "a theory that the differences between modern plants and animals are because of changes that happened by a natural process over a very long time." Google is the other go to source and quite arguably is used by more folks in need of a quick bit of information. Google chooses to display a definition of evolution as "the process by which different kinds of living organisms are thought to have developed and diversified from earlier forms during the history of the earth."

WHAT IS EVOLUTION?

Evolutionary biologists and researchers in the same field of study prefer a more scientific source when defining evolution. Yet, Webster and Google must be in the mix, so to speak, due to their quick and easy access to the general public. Unfortunately, the acquisition of scientific terminology pertaining to evolution is not so straightforward. Customarily speaking, each proponent of what made Darwin a celebrity, tends to craft a unique version of what they believe evolution to be.

Douglas J. Futuyma has a PhD in zoology, has taught as professor of Ecology and Evolution at Stony Brook University in New York, served as chair of the Department of Ecology and Evolutionary Biology at the University of Michigan, Ann Arbor, served as president of the American Institute of Biological Sciences, and is the author of many undergraduate textbooks on evolutionary biology. Futuyma defined evolution as, "change in the properties of populations of organisms that transcend the lifetime of a single individual. The ontogeny of an individual is not considered evolution; individual organisms do not evolve. The changes in populations that are considered evolutionary are those that are inheritable via the genetic material from one generation to the next. Biological evolution may be slight or substantial; it embraces everything from slight changes in the proportion of different alleles within a population (such as those deter-

WHAT IS EVOLUTION?

mining blood types) to the successive alterations that led from the earliest proto-organism to snails, bees, giraffes, and dandelions."

The author of *Why Evolution is True* and professor of Ecology and Evolution at the University of Chicago, Jerry A. Coyne defines evolution by saying it "can be summarized in a single (albeit slightly long) sentence: Life on earth evolved gradually beginning with one primitive species - perhaps a self-replicating molecule - that lived more than 3.5 billion years ago; it then branched out over time, throwing off many new and diverse species and the mechanism for most (but not all) of evolutionary change is natural selection."

Best known as television host *Bill Nye, the Science Guy*, Mr. Nye is also the author of *Undeniable: Evolution and the Science of Creation.* In his book he expressed that evolution takes place when "genes mutate enough through enough generations, and you get individuals that can no longer reproduce with each other; they've become separate or a new species."

Unbelievably, Darwin himself never defined the word "evolution" in his great work known as *The Origin of Species*, but did say that species are "mutable." Mutable species means that, with time, the genetic boundaries of a species can blur. This idea is the basis for what he called "descent

WHAT IS EVOLUTION?

with modification from a common ancestor," and is the engine that, according to evolutionary theory, caused the first unicellular organism to increase in complexity, branch off, and populate the Earth with seemingly countless species. Yet, does the idea of "descent with modification from a common ancestor" accurately describe evolution in totality? Could this statement also be used to describe the grandchildren/descendants of a matriarch that have differing appearances/modifications due to trait variation? Also, does "descent with modification" account for organisms that do not undergo modification? Why do lower forms of life still exist? There must be a more accurate description of evolution than what Darwin hinted toward.

The National Academy of Sciences was created by an "Act of Congress, signed by President Abraham Lincoln in 1863, the National Academy of Sciences is charged with providing independent, objective advice to the nation on matters related to science and technology." The National Academy of Sciences dissects evolution into three unique definitions:

"**Evolution:** consists of changes in the heritable traits of a population of organisms as successive generations replace one another. It is populations of organisms that evolve, not individual organisms."

"**Microevolution:** Changes in the traits of a group of organisms within a species that do not result in a new species."

"**Macroevolution:** Large-scale evolution occurring over geologic time that results in the formation of new species and broader taxonomic groups."

WHAT IS EVOLUTION?

Why does the National Academy of Sciences hold the position that evolution is a fact, yet fail to offer a single and concise definition?

In 1925, the first legal battle erupted when a science teacher in Tennessee taught evolution. The State of Tennessee v. John Thomas Scopes was a highly publicized case and regardless of the outcome, it vaulted the contention over origins and what should be taught into the public eye. Just shy of a century later and the struggle is still being played out in our nation's courtrooms, but what is the struggle about? Surely the courts must have a working definition for evolution? No sane person would enter into a legal battle over the teaching of a certain topic without first establishing a working definition that could be scientifically and legally agreed upon. Sadly, our judicial system has failed to define evolution, yet has been arguing over it for generations. Court transcript after court transcript has proven that our legal system has failed to establish a concise and scientifically agreed upon definition. If our court system has left evolution undefined all this time, what does this mean about the validity of past outcomes?

The United States values education and our next generation. We have created benchmarks for our students in the areas of science, technology, engineering, and mathematics. The federally funded Next Generation Science Standards (NGSS)

WHAT IS EVOLUTION?

is a national effort to create a framework for public school education when it comes to science. While expressing the importance of biological evolution to be included in science education, the NGSS has not defined evolution. Public schools are required to teach evolution, but the NGSS "are a set of science education standards, and do not prescribe a curriculum nor any vocabulary list with definitions. Definitions of evolution would come in curricula, which are chosen by individual states, districts, or teachers."

According to U.S. federal science standards, the individual states have say over how evolution is defined. In California, the state funded Department of Education (CDE) requires branch offices to oversee local school districts to ensure the California Next Generation Science Standards (CA NGSS) are being followed. Like the NGSS, the CDE has "not published an approved definition of biological evolution."

County offices across the state of California require local districts to teach evolution, but when the Los Angeles County Office of Education (LACOE) is asked to provided a definition of evolution the response given is that the "LACOE does not provide this kind of assistance and we typically do not provide language for these kinds of curricular questions."

WHAT IS EVOLUTION?

Los Angeles Unified School District (LAUSD) is the second largest school district in this country with approximately 700,000 students. LAUSD is required to teach evolution as part of its science framework, but the district does not have an official definition. Public schools across the country do not teach a scientifically agreed upon definition of biological evolution. Currently, science teachers are required to teach evolution, but use a definition that has been provided by the local school board, or just one they prefer. The United States has spent taxpayers' dollars in order to create science standards, but what is standard about the evolutionary education across this country?

Biological evolution is undefined. Science teachers have the freedom to define the term as they please, which undermines a national standard for scientific education. It is unjust for the NGSS to be required, but for it to offer no definition of required terms. It is clear to see that contention still surrounds the topic of biological evolution, how can those who oppose it be justly required to learn about an undefined term? Does it seem logical to require evolution as part of an educational framework and not define it? Evolution is such a contentiously misunderstood idea, we must not continue without a clear, concise, legal, and scientifically agreed upon definition.

WHAT IS EVOLUTION?

CREATING A DEFINITION: One way to acquire a single and robust definition of biological evolution could be to base it on parts of the previously mentioned. In the definitions provided by Webster and Google, there is acknowledgment of a current variety of living organisms that came to be quite diverse over time. Echoing this theme, Douglas J. Futuyma speaks of the "earliest proto-organism" successively experiencing genetic alterations until the complex and diverse life of today comes onto the scene. Jerry A. Coyne continues to expand the concept of life branching out and throwing off new and diverse species and also states that "natural selection" is part of the mechanics behind it. Bill Nye's definition highlights how gene mutation adds up over many generations to create new species that are no longer able to interbreed. Darwin was on the scene too early to have knowledge of the field of genetics, but displayed amazing foresight when stating his conclusion that species are "mutable." In the first definition offered by the National Academy of Sciences, it is made clear that "populations of organisms" evolve, not individuals.

Now that a recapitulation of the previously mentioned definitions of evolution has taken place, perhaps the following statement could serve well as a working model for the remainder of this volume.

WHAT IS EVOLUTION?

Evolution is the emergence of diverse, genetically isolated organisms from a common ancestor, by means of natural forces that select genetic mutations to be laterally and/or vertically transferred to successive generations.

RESULT AND MEANS: This new definition of biological evolution brings a greater level of understanding to the term because it clearly outlines its result and means. If all of life's biodiversity is the result of evolution then it must be stated up front, hence the first section of this new definition states "Evolution is the emergence of diverse, genetically isolated organisms." The National Academy of Sciences' definition of species speaks of how a "species consist of individuals that can interbreed with each other," with that said, a genetically isolated organism is unable of producing viable/fertile offspring with an organism of differing species. The section immediately following states where these new species originated which is "from a common ancestor." Clearly outlining evolution's result is very important and is done so in the first section: "Evolution is the emergence of diverse, genetically isolated organisms from a common ancestor." The reason we humans are on a quest to understand the enormous diversity of life on Earth, is the enormous diversity of life on Earth. Charles Darwin wrote *The Origin of Species*, because he wanted to explain the origin of diverse species. Evolution's life blood is the emergence of new and unique kinds of living things, and if speciation is not included throughout the conjecture, then those engaged are spuriously depicting evolution.

WHAT IS EVOLUTION?

Science is on the quest of "how", so when defining evolution the "how/means" must remain part of the quest. In the second section of this new definition, the mechanisms necessary for producing "diverse, genetically isolated organisms" are explained. Evolution happens "by means of natural forces that select genetic mutations." Darwin detailed beautifully in *The Origin of Species* how a breeder can select a desired trait among pigeons and make sure to only have certain males breed with certain females that possess said trait. This increases the occurrence of the selected trait in the offspring of the pair selected for breeding. Since breeder selection does not occur in nature, it is called artificial selection. Therefore the opposite is true as highlighted in the following hypothetical situation. If a drought causes some organisms to perish, yet some drought resistant organisms survive long enough to pass drought resistant genes to their offspring, a "selection" took place. The amount of rainfall a certain area receives is part of nature, so this kind of "selection" is called "natural selection." The second part of this new definition of evolution describes evolution as happening "by means of natural forces that select genetic mutations to be laterally and/or vertically transferred to successive generations."

Producing "successive generations" is the most crucial part of evolution, for in it holds the future of the species that have yet to emerge. If the first unicellular organism did not possess drought resistant genes and happened to experience a drought before it produced a "successive generation," all of life's diversity would have never existed. Genes must be transferred before an organism dies in order for the "selection" it went through to even matter evolutionarily speaking.

The last part of this new definition mentions natural forces selecting "genetic mutations to be laterally and/or vertically transferred to successive generations." Vertical gene transfer is the most familiar way of passing along heredity, for it is how we humans procreate. The passing of genetic material down to an offspring is a vertical gene transfer. The reason this new definition includes a second way to transfer genetic material has to do with studies that suggest certain types of unicellular organisms have the ability to transfer genes to each other laterally without creating a new individual. An easier way to grasp this concept could be similar to, yet very different from how a pre-existing infant builds up immunity and antibodies through breast milk. Antibodies or bits of biochemical information are transferred without producing a new individual. The mentioning of lateral gene transfer acknowledges that some unicellular organisms have the ability to give, share, or trade parts of their DNA, which is why this new definition speaks of two types of gene transfer.

WHAT IS EVOLUTION?

Evolution is the emergence of diverse, genetically isolated organisms from a common ancestor, by means of natural forces that select genetic mutations to be laterally and/or vertically transferred to successive generations.

An even more simplistic way of communicating this new definition could be:

Evolution is speciation caused by naturally occurring selective pressures.

The emergence of new species - genetically isolated organisms - is the result and so much of what evolution encompasses that Jerry A. Coyne wrote in his book *Why Evolution is True,* "if speciation didn't occur, there would be no biodiversity at all." The subsequent means of how the result came to be is just as imperative. Darwin was on the right train of thought when he gave a hint toward the idea of harmless mutation (genetic mistakes that do not decrease an organism's ability to survive) and genetic drift (the random sampling of traits/genetic information from two parents, in turn creating offspring with their own unique traits/genetic information) when he stated species to be "mutable." The two mechanisms of mutation and genetic drift, when combined with natural selection, are the evolutionary means of life's biodiversity.

Now that this new definition of evolution has been proposed, let us explore the pages of this volume in a more

WHAT IS EVOLUTION?

focused light. Hopefully, the confusion will end, the fog will rise, and the double talk will cease.

Chapter 2.

WHAT ARE SPORES OF CONFUSION?

"Evolution is within us, around us, between us, and its workings are embedded in the rocks of aeons past."
Richard Dawkins

The most common way evolutionary spores of confusion proliferate is through the interchangeable use of the word "evolution." Many subscribers of Neo-Darwinian thought use the word in a "catch all" kind of way and mold, shape, or stretch evolution to fit in place of domestication, artificial selection, natural selection, adaptation, trait variation, speciation, and the list goes on. Granted these processes are cogs in the machine of evolution, yet much like a transmission is part of an automobile; no one is calling it a car.

DOMESTICATION: Unfortunately, an interchangeable and all encompassing use of the word "evolution" is what many proponents of Neo-Darwinian thinking tend to communicate. Bill Nye equates evolution with the human ability to hybridize plants. This kind of thinking tends to bleed into most people's evolutionary view of what it means to breed dogs, horses, cattle, pigeons, fish, corn, wheat, tomatoes, et cetera. Breeders and growers have the ability to "artificially select" certain traits they wish to see flourish or subside in the genetic line of the plants and animals they are cultivating/breeding. This type of "artificial selection" speeds up and intensifies the effects of selection beyond what is seen in nature. Of course, we have seen astounding and drastic results from breeders who select dogs with polar opposite traits in size, yet Great Danes and Chihuahuas are both still dogs with the ability to interbreed. Despite the vast array of canine shapes and sizes, dog breeders are all dealing with one species. Two varying breeds have the ability to produce viable offspring which is quite telling; they are of the same species.

Ernst Mayr was a well respected ornithologist, taxonomist, and an evolutionary biologist. He is most famous for his part in developing the Biological Species Concept. In 1942, Mayr defined species as "a group of interbreeding natural populations that are reproductively isolated from other such groups." Mayr's definition means that organisms that produce viable offspring together are of the same species and organisms that do not produce viable offspring together are of differing species. This definition is currently the accepted standard that biologists use when studying organisms.

According to the National Academy of Sciences, selective breeding is the "intentional breeding of organisms with desirable traits in an attempt to produce offspring with enhanced characteristics or traits that humans consider desirable. This process is also known as 'artificial selection' (compare with 'natural selection')." This selective type of breeding that we humans have been doing for millennia, is not only possible among canines, but also very possible and fruitful when breeding cats, cows, corn and a whole host of other organisms. Our ability to identify traits of organisms and cause a species to be more useful is nothing new, for we have been building civilizations around this ability since the dawn of the neolithic era. Yet as of late, some Neo-Darwinian conversations have taken what is usually called domestication, and have used it interchangeably with the word "evolution." The practice of humans selecting desirable traits in other living things in order to gain a higher level of usefulness from said organism is called domestication via artificial selection. When the conversation turns toward the result of "breeding," we need to be honest with our words and use artificial selection and/or domestication.

WHAT IS EVOLUTION?

NATURAL SELECTION: The practice of a breeder selecting the traits of an organism to be passed on to the next generation is referred to as artificial selection and/or domestication. Natural selection, however, is when an organism is selected by nature to pass on traits to successive generations. Of course, nature does not intentionally select the way humans do, for organisms are selected by becoming a "lucky" gene pool that somehow avoids death from predators, natural disasters, and/or plain stupidity. Darwin described natural selection as being just like a breeder's selection, yet nature makes the selection. Much like artificial selection and/or domestication, natural selection is capable of causing great visual differences. Yet, this adaptation do to natural pressures is just that... adaptation.

ADAPTATION: In Darwin's book, *The Origin of Species*, he pointed out varying beak shapes and sizes of the ever so infamous finches of the Galapagos. He used their obtusely recognizable differences to infer the emergence of new species when outlining his ideas. Peter and Rosemary Grant are both Emeritus Professors in the department of Ecology and Evolutionary Biology at Princeton University. They spent four decades on the Galapagos Islands performing research in the evolutionary biology of finches. Their observations have shown that, regardless of beak shape and/or size, the varying finch inhabitants have been observed producing viable offspring with each other. Much like the previously mentioned example of a Great Dane's ability to interbreed with a Chihuahua, the equally noticeable differences in finch anatomy have not erased their ability to interbreed. If the finches of the Galapagos are able to produce viable offspring regardless of beak shape and size, is it logical to say new finch species have emerged?

The Grants realized that beak size variation among the Galapagos finch population is linked to the seasonal shift in food source availability; finches that have the most nutrients tend to have the most offspring. When food sources seasonally shift, one of the other finch beak shapes becomes more advantageous. Finches with the best ability to consume calories have been observed to dominate reproductively, which passes along traits that have an advantage when consuming the same food source. The shift in finch appearance seems to be an indicator of the emergence of a new species, yet the observation of finches producing viable offspring between visually different finches establishes otherwise. Also, the genetic sequencing of varying Galapagos finch populations has confirmed the sharing of identical genes and their singular species. So, why does the appearance of varying species exist? Although the finches of the Galapagos have the same variance in DNA that any other gene pool would have, there are two genes that give rise to the variety of beak shape and size. These genes are regulated by switches that are very similar to a dimmer switch in one's living room. Research has shown finches of the Galapagos to have elongated beaks when their *CaM* gene is turned up to a higher level during development. In contrast, broader and deeper beaks form when the *Bmp4* gene is turned to a higher level at earlier stages in development. Geneticists have acquired much empirical data that displays beak shape and size to be determined by the timing and intensity of these genes being activated.

Changes in a finch population that are attributed to environmental factors like food source availability is called "adaptation." In fact, this shift is also common in many other species of birds, fish, plants, mammals, and insects. Humans, for example, have quite a large spectrum of visual differences which, in our horrific past, caused some to classify varying people groups as varying species. Yet, now we know that humankind is only one species. All ethnic groups among humans share the same genome and have the ability to interbreed and produce viable offspring.

Spores of confusion proliferate around adaptation more than any other term due to the inference of visual dissimilarities being an indicator of new species. Yet, empirical data has shown that many species that have been

WHAT ARE SPORES OF CONFUSION?

visually classified as unique are actually able to interbreed and produce viable offspring. Is it scientifically accurate to claim two separate species exist if a viable offspring is able to be produced? Many who are engaged in the conversation of origins use the word "evolution" in place of "adaptation" and this must be acknowledged. Drastic visual adaptations are acquired when organisms experience natural and/or artificial selective pressures, yet empirical data has not shown adaptations to cause the emergence of new species.

TRAIT VARIATION: Trait variation is the polite guest that is "paying for the party." If Darwin's finches did not have varying traits, there would be only one beak shape and size. This lack of trait variation would be quite negative, for just one seasonal shift in food source availability could very well decimate the Galapagos finch population all together. Fortunately, trait variation among these birds is a reality and has given nature something to "select."

Among most organisms, trait variation is very high. In nature, organisms usually do not experience strong enough selective pressures in order to display their unseen traits. Case in point, the wolf displays very little trait variation in the wild. Perhaps one pack may be larger or darker in color than other packs, but for the most part wolves are the same. Yet, look what artificial selective pressures have exhumed from the genetic coffers of these wild dogs. The genetic sequencing of various breeds has shown that all dogs were bred from wolf stock. Natural pressures were not able to reveal the extraordinary trait variation of the wolf, but once humans understood domestication, unseen traits came to the surface in all of the fascinating breeds we have today.

Trait variation and what organisms have at their disposal in the process of adaptation is extraordinary, and it is the polite guest that is "paying for the party."

SPECIATION: In his book *The Origin of Species*, Darwin wrote, "from so simple a beginning endless forms most beautiful and most wonderful have been, and are being evolved." He communicated that one species became many diverse species which he called "descent with modification from a common ancestor." He wrote of how mutation is the engine of life's diversity. His idea that species are "mutable" is at the heart of his work because it refers to an ability of one life form or species to mutate into another. This is speciation and it is the result of evolution. Lamentably, Darwin did not give a clear explanation of how "mutable" species mutate. Jerry A. Coyne wrote that a better title for Darwin's book, *The Origin of Species* would have been "The Origin of Adaptations: while Darwin did figure out how and why a single species changes over time (largely by natural selection), he never explained how one species splits in two.

WHAT ARE SPORES OF CONFUSION?

Yet in many ways this problem of splitting is just as important as understanding how a single species evolves. After all, the diversity of nature encompasses millions of species, each with its own unique set of traits. And all of this diversity came from a single ancient ancestor. If we want to explain biodiversity, then, we have to do more than explain how new traits arise - we must also explain how new species arise. For if speciation didn't occur, there would be no biodiversity at all - only a single, long-evolved descendant of that first species."

Does it seem logical to use the word "evolution" if it is not intertwined with ideas that reflect life's diversity? Is it scientifically accurate to interchangeably use other terms like domestication, artificial selection, natural selection, adaptation, and trait variation in the place of the word "evolution?"

Chapter 3.

IS EVIDENCE EVIDENT?

"The events that led from the first living cell to you and me have required a nearly unimaginable period of time. When we're talking about evolution, the expression 'a long time' is an understatement."
Bill Nye

The proposed humble origin of all multi-cellular life evolving from a primordial unicellular ancestor had to have been a very long time ago. The evolutionary smoking gun has completely off gassed and there are no eye witness accounts to speak of. There is much evidence that is available for study, but is evidence evident?

Time and time again, the bickering from every fledgling and well-established camp volleys a proclamation steeped in its own take of evidential validity. No one piece of evidence can be, nor has been, interpreted in one way. Examining evidence is like shining light, but shining light creates shadows. Moving the light to another aspect of evidence illuminates the shadows, but causes shadows to darken what was once illuminated. If the shadow shifting continued, when would it ever end? There is an enormous need for a final word to be spoken over a piece of evidence,

WHAT IS EVOLUTION?

but whose word is final? When evidence does not seem to be evident to all parties involved and the conjecture becomes seemingly endless, there must be a voice ready to say one way or another.

In science, the final word is experimentation. An experiment is comprised of empirical observation, testing, and repetition. Once evidence is discovered, there are strict guidelines in place that, if followed, would cause all to see the proposed explanation to be evident. Scientific evidence is up for conjecture, but can be validated through the shadowless and all encompassing light of an experiment.

After years of research, observation, wind tunnel analysis, and unmanned trials, the Wright brothers were ready to test their idea through unbiased and unforgiving experimentation. No longer would powered human flight be an idea with convincing evidence, for the brothers put their claims to the test. On December 17th, 1903, the Wright brothers were successful and the possibility of powered human flight was no longer up for debate. The past was chock full of attempts to show the worthiness of belief in their proposed hypothesis, but after that fateful day, acceptance was no longer necessary. One experiment validated the words they had spoken and powered human flight not only caused radical progress and innovation, but its possibility was made evident to everyone. The encompassing light of their experiment eradicated the shifting shadows of evidential interpretation. The Wright brothers seeded the expansion of scientific horizons and their inventions led to other aeronautic breakthroughs. Other scientifically focused minds built on their accomplishments and in sixty-six short years, the first human walked on the Moon.

Author of *Darwin's Dangerous Idea* and Co-Director of the Center for Cognitive Studies at Tufts University in Massachusetts, Daniel C. Dennett wrote about Darwin saying that he had "the greatest idea that anyone has ever had." If this statement is true and Darwin did indeed have an idea so great, the evidence for evolution must be evident.

The National Academy of Sciences holds the position that, "evolution is both a <u>fact</u> and a process that accounts for the diversity of life on Earth." Is evolution a fact? Does evolution have the ability to account for biodiversity? Let us take another step in this journey and refrain from shadow shifting by bathing the evidence in all encompassing light.

DIVERSITY OF SPECIES: As far as we humans can tell, Earth is the only dynamic host of animate beings; a real smorgasbord of biological diversity. Species after species have been discovered and we are still learning about new and exciting places where many, yet to be discovered, species live. Our humble blue-green oblate spheroid is no more than a pinhead in comparison to real estate on a universal scale, but is an actual colossus when it comes to life. According to current empirical data, nowhere else but Earth has life and nowhere on Earth is totally lifeless. Perhaps flowing lava sterilizes an area temporarily, but when it cools, rich nutrients are left behind that spur life on even more. There is so much life here that biologists find giving a definitive number of living species quite problematic. In fact, this great number of species is used as evidence for one of the leading trains of Neo-Darwinian thought. The idea is that all of the unique and abundant forms of life are a result of evolution. Bill Nye concludes that "biodiversity can be quantified," he claims it to be "the measure of the results of evolution."

Richard Dawkins is an evolutionary biologist that held a position at the University of Oxford as Professor for Public Understanding of Science, he is the founder of the Richard Dawkins Foundation for Reason & Science, and a renowned author of many books. In *The Greatest Show On Earth: The Evidence for Evolution* Dawkins wrote, "We are surrounded by endless forms, most beautiful and most wonderful, and it is no accident, but the direct consequence of evolution by non-random natural selection - the only game in town, the greatest show on Earth."

IS EVIDENCE EVIDENT?

At this juncture, the question of how these two and many others have come to this conclusion must be asked. What has logically and scientifically caused some to use life's diversity as evidence for evolution? Granted, there are many diverse, genetically isolated groups of organisms on this planet, which is a fact that is not up for debate. The means of how Earth's biodiversity emerged is in need of the all encompassing light of the experiment. Observation confirms that life is abundantly rich. Yet, what experiment could verify evolution to be the means of diverse species?

Is the existence of diverse species valid evidence for the means of how diverse species came to be? Is this argument based in science, or is this argument somewhat circular? If evolution were to be the means of Earth's diverse species; what supportive evidence could be tested? When it comes to the conjecture surrounding the origin of life's biodiversity, this is the question of most importance. If evolution is a fact and all species emerged from a common ancestor via natural forces, what is the supporting evidence for this idea and how can we bathe it in the all encompassing light of an experiment?

THE FOSSIL RECORD: Stone masons and lignite minors of antiquity would find bones of what they thought were giants and speculated great battles must have occurred. The pre-paleontological minds of the ancients had little to no understanding of fossilization, of what had been preserved, nor of when these mysterious bones had come to their final resting place. Over vast swaths of human history, we have burrowed and bored into just about anything in hope of unearthing precious metals, salt, fuel, gems, food, and buried treasure. In our quest for resources, great is the probability that hundreds or even thousands of fossils had been discovered. The ancient Greeks were the first to have documented mysterious ancient remains, but more than likely they were far from the first to have uncovered a fossil. Other cultures before the Greeks may have discovered fossils, but either paid no attention or chose not to document them. Perhaps some people groups had an aversion to the handling of remains and when fossils were brushed up against in the dredging process, they would backfill as quickly as possible and move locations. Maybe a certain culture was familiar with large extant creatures like elephants, giraffes, or whales and could easily dismiss strange bones as recognizable. Possibly the most common reason for why mankind did not document the discovery of fossils until Grecian times revolves around limited resources. Whether hunger or a neighboring hostile tribe were to blame, anything beyond the support of moment to moment survival might well have seemed nonessential. This hypothesis could give clarity as to why and when the explosion of interest in paleontology took place.

IS EVIDENCE EVIDENT?

In the mid to late eighteenth century, colonies had been settled across the globe. Some societies were finding themselves becoming very good at shipping, trade, warfare, mechanization, and documentation. The printing press had been available for some time and global correspondence between varying interest groups began to increase. Resources from the earth became commodities that were prized on a larger market increasing the potential of unearthing strange and mysterious bones. At this point in time, humans found themselves in a modernizing society with an increasing ability to devote time to inquiry and to confer with other intellectuals. Some intellectuals were finding themselves free of the shackles which had previously bound them to the support of moment to moment survival. The once nonessential ancient remains became a target of interest for the "well to do," giving much food for thought and fuel for conjecture. Society founded themselves a field of inquiry that could only be afforded to those no longer on the quest of daily survival. This new field of inquiry became less about progress and innovation and more about tickling the need some had in regard to pondering the "maybe" and the "what if."

The great and ingenious Benjamin Franklin was a thinker in his own right, and he too took a bit of time to indulge his curiosity in what he called, "Prehistoric Animals of the

WHAT IS EVOLUTION?

Ohio." Franklin wrote to a friend in 1768 about a fascinating discovery:

"[T]he skeletons of near 30 large animals suppos'd to be elephants, several tusks like those of elephants, being found with these grinder teeth - four of these grinders were sent me by the gentleman who brought them from the Ohio to New York, together with 4 tusks, one of which is 6 feet long & in the thickest part near 6 inches diameter, and also one of the vertebrae - My Lord Shelbourn receiv'd at the same time 3 or four of them with a jaw bone & one or two grinders remaining in it. Some of our naturalists here, however, contend, that these are not the grinders of elephants but of some carnivorous animal unknown, because such knobs or prominences on the face of the tooth are not to be found on those of elephants, and only as they say, on those of carnivorous animals. But it appears to me that animals capable of carrying such large & heavy tusks, must themselves be large creatures, too bulky to have the activity necessary for pursuing and taking prey; and therefore I am inclin'd to think those knobs are only a small variety, animals of the same kind and name often differing more materially, and that those knobs might be as useful to grind the small branches of trees, as to chaw flesh - However I should be glad to have your opinion, and to know from you whether any of the kind have been found in Siberia."

Benjamin Franklin may have continued his appeasement of curiosity in the "Prehistoric Animals of the Ohio," but no other letters have been found. Interestingly enough, this great man of invention, innovation, and progress was utterly cognizant of the obvious contention and degrees of opinion surrounding prehistoric life forms. Why is there no follow up letter to this conversation? Why is Benjamin Franklin not celebrated as our nation's first paleontologist? Could it have to do with the fruit of true scientific inquiry? When it comes to Franklin and his type of curiosity, he would generally revel in a great deal of meaningful result. Beside his apparent interest and much fruit revolving around his immersion in our young government, Franklin was a scientist. Just to name a few, he had experimented with electricity, conductors, hybridization of corn, hot-air balloons, the aerial voyage of man, and mathematics. Also, he charted the Gulf Stream when collecting data on his multiple voyages across the Atlantic. Benjamin Franklin not only ran a lucrative printing business, but was a pioneer in counterfeit-proof paper money. A few of the inventions attributed to him are the lightning rod, bifocals, daylight savings, and a more efficient type of fireplace. Franklin was a curious man, but he put his curiosity to work and expected much valuable fruit to transpire. We will never know why Franklin only wrote one letter on the subject of the "Prehistoric Animals of the Ohio," but perhaps his interest in paleontology decreased because he realized that it is a field of inquiry which does not envelope the raw materials which lead to scientific breakthrough. If Benjamin Franklin were able to visit a natural history museum today, he might believe that viewing the display of ancient remains is no more scientific than visiting an art gallery. Much like the subject in Leonardo da Vinci's Mona Lisa, organisms that have been captured in stone are unable to be scientifically observed. Of course, science can be done to discover the compounds and pigments da Vinci used, but nothing about his subject can be observed other than her image. Fossils can be scientifically studied when it comes to observing and testing the composition of the rock, yet the organism preserved can also only give an image to observe. Behavior and motor skills can not be witnessed. DNA can not be sampled. Diet and sleep patterns can not be studied. Beside an organism's shape, size, and location of discovery, what observation, and empirical data can be collected? When it comes to the study of the remains of fossilized beasts, inference and postulation tend to be the tools of most importance. We will never know why Benjamin Franklin is not celebrated as our nation's first paleontologist, yet we do know that he was a man of true scientific inquiry which allowed him to revel in a great deal of meaningful result.

What can the fossil record tell us about evolution? In *The Origin of Species*, Darwin states:

"[T]he extermination of an infinitude of connecting links, between the living and extinct inhabitants of the world, and at each successive period between the extinct and still older species, why is not every geological formation charged with such links? Why does not every collection of fossil remains afford plain evidence of the gradation and mutation of the forms of life? We meet with no such evidence, and this is the most obvious and forcible of the many objections which may be urged against my theory. Why, again, do whole groups of allied species appear, though certainly they often falsely appear, to have come in suddenly on the several geological stages? Why do we not find great piles of strata beneath the Silurian system, stored with the remains of the progenitors of the Silurian groups of fossils? For certainly on my theory such strata must somewhere have been deposited at these ancient and utterly unknown epochs in the world's history."

In his day, the fossil record placed Darwin at a loss and he was quite humble about it. Today, many evolutionary biologists have similar things to say about the fossil record and seem to be facing a century and a half old problem. Jerry A. Coyne wrote that "for every two species, however different, there was once a single species that was the ancestor of both. We could call this one species the 'missing link'. As we've seen, the chance of finding that single ancestral species in the fossil record is almost zero. The fossil record is simply too spotty to expect that." Bill Nye agrees with Darwin and Jerry A. Coyne by saying, "we don't see the intermediate linking individuals, because there were so few of them." What evidence for evolution does the fossil record offer? Why do we not find a fossilized, gradual transition from species to species? Could there be an explanation for the sudden appearance of fully formed organisms in the fossil record?

The foundation of evolutionary theory is the idea that mutations take place over long periods of time and gradually produce new life forms, yet in our natural world today, new species are not observably emerging. Evolutionary biologists have turned to the fossil record to find evidence to support foundational ideas, but have realized there is a lack of representation. What is represented in the fossil record are major and abrupt additions of new species and phyla (large classifications of organisms) in their fully developed forms. When speaking of this issue, Bill Nye stated that once you "understand genetic island formation or punctuated equilibrium, it would be weird if things were any other way. The missing nature of missing links is actually further proof of evolution."

Is Bill Nye using sound logic when using "the missing nature of missing links" as evolutionary proof? What else could be verified using this type of thinking? Could the fact that no one has ever seen, fill in the blank, be proof that it exists? Does Mr. Nye's mentioning of "the missing nature of missing links" as evolutionary support prove to be consistent with empirical methodology? Would understanding the intricacies of "genetic island formation" and "punctuated equilibrium" truly give understanding? Perhaps now would be a good time to stop and see if the unpacking of two additional theories will cause the original theory to become self-evident.

PUNCTUATED EQUILIBRIUM: Today, biologists struggle with the same issue of transitional fossils of which Darwin himself struggled. An explanation has been created for this problem and it is called punctuated equilibrium. Stephen Jay Gould was an American paleontologist that put much time and energy into the American Museum of Natural History in New York, he was a Harvard professor, and the co-creator of a theory called punctuated equilibrium. Gould and his co-creator wrote that the "history of evolution is not one of stately unfolding, but a story of homeostatic equilibria, disturbed only 'rarely' (i.e., rather often in the fullness of time) by rapid and episodic events of speciation."

This theory was created to explain the abrupt appearance of fully formed species in the fossil record. Punctuated equilibrium predicts evolutionary events to have taken place far too quickly to have been documented in the fossil record, thus explaining the sudden appearance of new and fully formed fossilized species. In layman's terms, punctuated equilibrium explains that we do not see new species gradually appearing in the fossil record because species emerge faster than the amount of time needed for fossilization to take place.

Is it logical to claim that speciation happens faster than fossilization, yet also claim the emergence of species takes place over millions of years? Is true science being represented when evolutionary biologists, not only promote the idea of punctuated equilibrium, but also state that evolution unfolds over eons? When speaking of evolution, biologist claim that speciation could have taken millions of years or it could have happened so quickly that the fossil record was not able to document the change. Does this type of rationale confirm that empirical data and logic are paramount to evolutionary thought?

WHAT IS EVOLUTION?

GENETIC ISLAND FORMATION: The second idea Bill Nye suggests that must be understood in order to fully grasp evolution is "genetic island formation," otherwise known as speciation. When a group of organisms are on a genetic island, it means that said organisms are unable to transfer genes with organisms outside of the group. No one is certain how genetic island formation takes place, but most believe "allopatric speciation" to be the best explanation. "Allopatric" is a Greek word comprised of "allo" which means "other" and "patric" means "fatherland." In essence, a group of organisms might find themselves split in two by a geographic separation. For example, one way allopatric speciation may occur is for a group of land dwelling organisms to be separated in a storm. Perhaps some of the group members drift to a neighboring island on a tree trunk and then begin to colonize. The population of interbreeding organisms is separated geographically and therefore becomes unable to interbreed with the whole population due to separate localities. Time goes by, mutations and genetic drift add up which causes the two groups to adapt and change at different rates due to differing selective pressures. In the future, ocean levels may drop creating a land bridge or perhaps tectonic forces bring these two populations together again; the possibilities are endless. The idea is that once the previously separated group of a single species are reunited via geographic barriers no longer being a factor, the two are supposedly unable to interbreed due to the accumulation of genetic mutations. The two groups, in theory, have become two unique species. What evidence has been found to support this idea, and how can it be bathed in the all encompassing light of an experiment? What does empirical data confirm?

Allopatric speciation (the emergence of diverse species via geographic separation) is the best explanation evolution has for life's biodiversity. Yet, biologists today are collecting data that suggests the opposite. According to research performed by San Diego State University, biologists have noticed that geographic separation and habitat "fragmentation can lower migration rates and genetic connectivity among remaining populations of native species, reducing genetic variability and increasing extinction risk." The California State Parks system released similar information that said, "From a land-borne species perspective, these wildland remnants end up trapping some species with little or no available resources for survival, which can lead to extirpation (extinction)."

In evolutionary circles, the term "bottleneck" is used to describe the process of a population moving through a reduction of members. This reduction could come through geographic separation, natural disaster, disease, over hunting/fishing, or any cause that would "thin out the herd" so to speak. A reduced population is a reduced gene pool which is a very important part of the evolutionary founder effect of a new colony. Species that have become geographically isolated in a reduced gene pool are inferred to have eventually become genetically isolated, emerging as a new species. According to Neo-Darwinian hypotheses, genetic bottlenecks are the catalyst for speciation, but what does the data say?

The University of California at Berkeley communicates that "small populations face two dangers - inbreeding depression and low genetic variation... [for example] a population of 40 adders (Vipera berus) experienced [an] inbreeding depression when farming activities in Sweden isolated them from other adder populations. Higher proportions of stillborn and deformed offspring were born in the isolated population than in the larger populations... For Swedish adders, the solution to the inbreeding depression problem was simple - introduce adders from other populations."

Empirical data shows inbreeding to be far from beneficial and in most cases will cause a smaller, separated population to go extinct. Also, there is no data to suggest that populations would become genetically isolated from the original group if divided. For in the Swedish countryside, the introduction of adders from another population showed to be reproductively beneficial, highlighting the lack of speciation among the bottlenecked population. If observation tells us that geographic separation of species leads to an inbreeding depression and inbreeding leads to many negative effects and extinction, why would the idea of "genetic island formation" seem plausible? Could inbreeding have produced a positive result in the distant past? When a bottleneck or reduction in a population is observed today, empirical data shows it to be very detrimental for the affected population. Is it logical or even scientific to claim allopatric speciation to be the cause of genetic island formation if supportive observation and data cannot be collect in the present? Was Bill Nye correct in saying that the nature of missing links would become clear once we understood punctuated equilibrium and genetic island formation?

LIVING FOSSILS: Understanding evolution can be confusing, especially when there are so many proposed organisms that must have existed, yet are missing from the fossil record. Ever since Darwin published his great work, *The Origin of Species*, naturalists across the globe have been clamoring to fill the gap in regard to intermediary species. As previously mentioned, evolutionary biologists are unable to scientifically explain how the lack of transitional fossils supports evolution. With that said, what scientific explanation accounts for the innumerous fossils displaying no evolution whatsoever? Over millions and sometimes

WHAT IS EVOLUTION?

billions of years, a multitude of organisms have been captured in stone, yet populations of their species are still with us today. How could this be? How could evolutionary pressures be so powerful to have produced all of life's diversity, and at the same time produce zero evolutionary change? Why are there species in every major category of life that have remained in evolutionary stasis (unchanged) since their first appearance in the fossil record? According to the theory, evolution takes place no matter what the environment is doing due to genetic drift. How is it possible for organisms alive today to also be represented in the rocks of eons past? Even if an organism could find itself in a stable environment void of changing natural selective pressures, would not the random sampling of genetic information from two parents, in turn creating offspring with their own unique genetic information have occurred over and over until the ancestors' descendants became something quite different? Richard Dawkins gives a lovely description of how a group of organisms drift into an entirely different organism. In his book *The Greatest Show On Earth: The Evidence for Evolution*, Dawkins sets up the following "Thought Experiment."

"Slowly, by imperceptible degrees, the shrew-like animals will change, through intermediates that might not resemble any modern animal much but strongly resemble each other, perhaps passing through vaguely stoat-like intermediates, until eventually, without ever noticing an abrupt change of any kind, we arrive at a leopard."

IS EVIDENCE EVIDENT?

Richard Dawkins is very good at communicating what evolutionary theory predicts and what some evolutionary biologists postulate to have happened, but what does the data show? Does the fossil record reflect a slow and imperceptible change to organisms that pass through various intermediate forms? How logical is it to say that organisms drift through time and are molded into something new, even though we have living fossils that verify the opposite of this claim? Does evolution make a valid explanation for extant organisms that have displayed no evolutionary change in comparison to their fossilized counterparts? How could evolution not have affected bats, turtles, frogs, alligators, crocodiles, dragonflies, bees, mosquitoes, opossums, cycads, koalas, platypuses, cockroaches, chevrotains, lungfish, nautiluses, crinoids, horseshoe crabs, ferns, ctenophores, stromatolites, elephant sharks, pygmy right whales, ginkgo biloba trees, coelacanth fish, wollemi pines, many types of bacteria, microfossils, and more? (To view a non-exhaustive list of extant organisms and their fossilized counterparts, visit: http://tinyurl.com/Living-Fossil-List) Dare it be asked, are all "living fossils" immune to evolution? How could millions of years have gone by and left a multitude of living organisms unchanged? Why do so many fossils of the extant exist? Why are there so few transitional fossils? If observation tells us that evolution has not occurred among living fossils, how logical is it to infer that evolution took place among the life forms we can not observe?

The fossil record is an amazingly breathtaking natural phenomenon. Countless lifeforms have seen their last day when frozen in time, but what do fossils confirm other than organisms had once lived and died? Gorgeous details have been preserved in stone and are now awaiting discovery. Yet, every fossil that has been unearthed is of an organism still represented in life's diversity today, or of one that has gone extinct. If we can observe extinction and also observe how evolution has not affected observable "living fossils," how logical is it to infer that a transitional ancestral species must have existed even if it is not represented in the fossil record?

HOMOLOGY: Various sources for nomenclature in regard to the English language are derived from the Greek. Such is the case for the study of corresponding visual similarities between species, which is otherwise known as homology. The Greek word "homologos" split in two is "homo," meaning "the same" and "logos," meaning "related to." In the evolutionary field of study, homology is the comparison of visual, embryonic, and genetic similarities between species which are then inferred to confirm species' common ancestry.

IS EVIDENCE EVIDENT?

The most common avenue to the homologous approach revolves around visual comparisons of structural similarities found among skeletons of the extant and the fossils of the extinct. One can quickly see an apparent symmetry to mammals, for example; a bat has corresponding wing bones to a rabbit's corresponding front leg bones, and both seem to have an echoing presence in the front legs of a deer. In fact, a frog and a Tyrannosaurus Rex also have corresponding front leg and arm bones that reflect the possibility of common ancestry with mammals. The loudest proponents of descent with modification agree that mammalian evolution began with a small shrew-like creature that has been called Eomaia. It is positively evident to see that, other than its wings, bats and shrews have a strikingly comparative homology. Basing the validity of evolution on structural homology, one should be able to trace obvious mutations through the fossil record. If in deed, shrew-like mammals (Eomaia) were the ancestors of bats (which evolutionary theory predicts), then intermediate species must have been fossilized in the process. Disastrously, no such specimens have been unearthed. As previously covered in the section entitled THE FOSSIL RECORD, this problem of missing intermediate species is no new problem. Darwin, himself, was honest about the lack of evolutionary support in the area of transitional fossils.

WHAT IS EVOLUTION?

Visually identifying homologous similarities becomes dubious for two major reasons. The first has to do with our human bias. Author of *Cognitive Psychology: Connecting Mind, Research and Everyday Experience*, E. Bruce Goldstein who earned his PhD from Brown University and is also a graduate professor in the Department of Psychology at the University of Pittsburgh wrote about confirmation bias saying it is, "the tendency to selectively look for information that conforms to a hypothesis and to overlook information that argues against it." If an evolutionary biologist studies a fossil and already has a preconceived notion of evolution being a fact, how would that affect their ability to be objective? Is it possible that dissimilar features would go unmentioned while features that are comparative would become overly highlighted? As emotional attachments to a hypothesis increase, what is the likelihood of our propensity toward empirical methodology subsequently decreasing? Evolutionary biologists tend to claim Neo-Darwinian explanations to be fact before the analysis of new fossils begins. Is it possible to make visual comparisons that are free of confirmation bias if the hypothesis in question is touted as fact before research begins?

The second reason why identifying homologous similarities becomes dubious has to do with striking resemblances between one phylum to another. How must we reconcile the homologous attributes of organisms that are visually similar, but are not considered to be closely related? Why are visual similarities between sharks and dolphins not attributed to confirming recent common ancestry? How could it be that cuttlefish and macaws have identical beaks? Why do the color changing chromatophores in a chameleon's skin also exists in cephalopods? If land animals evolved from sea life, why don't all land animals (not just those of us higher up on the tree of life) have complex eyes like that of an octopus? When it comes to homologous similarities that are beyond the bounds of a phylum, why is the possibility of common ancestry excluded? Why are relational comparisons drawn between winged creatures of like phylum, yet the possibility of common ancestry is excluded from winged creatures of varying phyla? Why do some choose to say that organisms of obtusely different classifications that possess homologous features do not share a recent common ancestor? When do structural similarities no longer infer relational ties from one organism to another? Who gets to say that one similar feature among organisms does in fact confirm recent common ancestry, but a similar feature between members of varying phyla does not? (to view varying phyla with homologous adaptations, visit: http://tinyurl.com/Convergent-List)

Sir Gavin de Beer was an evolutionary embryologist, Fellow of the British Royal Society, and served as Director of the

WHAT IS EVOLUTION?

British Museum of Natural History. He has done extensive research on the embryonic development between organisms that are visually and structurally homologous. His experiments were designed to trace the development of certain attributes from the moment of fertilization until full development. The results of his work revealed that fully formed corresponding homologies do not develop from the same embryonic location. Sir Gavin de Beer realized that similar attributes develop from a fertilized egg in different locations, depending on the species. This means that if homologous structures do not develop from the same embryonic location, then they are not genetically related because their development is not controlled by their homologous (similar) DNA. In short, embryonic homology promptly discounts any possible attempt of supporting evolution via genetic homology. Sir Gavin de Beer said it best when saying, "It is now clear that the pride with which it was assumed that the inheritance of homologous structures from a common ancestor explained homology was misplaced; for such inheritance cannot be ascribed to identity of genes."

Homology makes visual, embryonic, and genetic comparisons, yet which of these is "evident" evidence for evolution?

IS EVIDENCE EVIDENT?

THE MYTHICAL NEANDERTHAL: Deoxyribonucleic acid (DNA) is a molecular instruction manual from which all living things are run and developed. Every germinated seed and growing embryo is taken through a step-by-step process of maturation via the detailed information stored in its DNA. Life is bursting at the seams with enumerated variation and each living thing displays a copious amount of unique detail. There is so much information in DNA that just one gram is capable of storing approximately 700 terabytes of genetic code. Researchers have mapped the human genome and counted the base pairs, but still have a rudimentary understanding of how it functions. DNA is still a mystery in many ways.

Some in the field of evolutionary biology believe that homologous DNA is powerful evidence for species mutating into new species due to the great similarities found in all organisms. The work of Sir Gavin de Beer has verified through observation that genetic similarities do not confirm common ancestry of varying species. So, what can DNA tell us about the validity of evolution?

Today, paternity tests are commonplace. Geneticists have the ability to test the DNA of certain humans (same species) in question and confirm a child's father. It is also possible to have an individual's DNA sequenced in order to discover their country or people group of origin. This ability has

WHAT IS EVOLUTION?

opened a new and exciting field of which much fruit has transpired in what is called the Human Genome Project. The human species is vastly different on a genetic level and those differences can be seen in our DNA. It is possible for geneticists to give an individual their unique percentages of European, African, Asian, and Native American ancestry. Not surprising, evolutionary geneticists have begun to use this technology to map our genetic road to becoming human. Recently, DNA extracted from Neanderthal bones was sequenced and has revealed shocking news. Geneticists have discovered that most non-Africans alive today have what is being called "Neanderthal genes" represented in their DNA. This discovery is fascinating, yet how is this possible?

Take, for example, wolves and varying breeds of domesticated dogs. We are able to observe them alive today and the conclusions we make are based on empirical data. We know that wolves and dogs are very different, but observation and DNA sequencing has shown they are able to produce viable/fertile offspring, which according to our understanding of biology, confirms a singular species.

When it comes to humans and Neanderthals, we are limited. We can not observe Neanderthals and collect empirical data that is free of inference. DNA is what can be compared. We can be positive that DNA sequenced from humans is in fact from humans, but we can only hope the DNA that has been sequenced from Neanderthal bone samples is actually from Neanderthals.

Unlike the wolf, domesticated dogs, and humans; no one has ever observed living Neanderthals. There are many dig sites, and much evidence to suggest the existence of a human/human-like group that could be inferred to have been Neanderthals, but we are not certain. Many scientifically minded researchers have not stopped at inference, but have continued to search for empirical data. Research has shown the human genome to possess similar sequences to that of the proposed Neanderthal genome. This data is so compelling that geneticists have arrived at the consensus of believing Neanderthals and humans to have interbred, producing viable/fertile offspring. Which according to our understanding of biology means the two groups were/are one species.

Although the idea of Neanderthals is fascinating, the empirical data suggests their DNA to be human. The most scientific way to value the empirical data given to us by the proposed Neanderthal, is to view this group of humans as a unique ethnicity. Much like a four foot tall African Pygmy tribesman could produce viable/fertile offspring with a six foot tall Swedish Supermodel, the differences are only ethnic and not a matter of species.

WHAT IS EVOLUTION?

The National Academy of Sciences defines species as a group of "individuals that can interbreed with each other." Mules and Ligers can make this definition seem confusing, but they are in fact hybrids of two separate species. They are sterile and can not go on to produce viable/fertile offspring. According to genetics, humans have bred with Neanderthals and have had viable/fertile offspring. Does this mean humans and Neanderthals are in fact one species? If Neanderthals and humans have produced viable offspring, how logical is it to say Neanderthals ever existed? Does empirical data show us that Neanderthals were a unique species or a unique ethnicity?

The science and data behind the study of the human genome has shown to be quite enlightening when it comes to our ancestry. If it is safe to say that all specimens classified as Neanderthal are in fact human, what about other early-human specimens like "Lucy" the Australopithecus afarensis? Observations have shown us that human skull variation today is no less extreme than skull variation found among complete fossils of early-humans. In Dmanisi, Georgia cave excavators have exhumed the remains of humans that lived within years of each other. The skull variation between each specimen is so extreme that if they had been found in different locations around the globe, researchers would have classified them as distinctly different species separated by eons. This discovery raises many valid questions in regard to all early-human remains and the classification of their separate species. One thing that can shed much light on early-human classification is the data given to us through genetic sequencing. "Lucy" is classified as a member of the Australopithecus species, yet her DNA is no longer viable for sequencing. This leads us to a very interesting question. If the remains of the Neanderthal contain viably sequenceable genetic material and are showing their oneness of species with humans, why then would we assume other early-human remains like "Lucy," Homo erectus, and Homo habilis, of which do not contain viably sequenceable genetic material, to be a unique species? How logical is it to have empirical data that clearly shows the ability of Neanderthals to produce viable offspring with humans, yet disregard this data when inferring the existence of other early-human species? Is it scientifically logical to claim the Neanderthal, "Lucy," Homo erectus, and Homo habilis to be unique species? If so, what empirical data suggests this to be true?

MUTATION: Another way DNA is used as evidence in the support of Neo-Darwinian conclusions has to do with the idea that random mutations accumulate and over epoch after epoch new descending species are churned out. Is this a possibility, could life have evolved over millions of years due to the accruance of genetic mutations?

WHAT IS EVOLUTION?

Genetic mutations are not to be confused with the random sampling of DNA from two parents to their offspring. Many in the field of evolution claim trait variation to be an example of genetic mutation, but that is a spurious depiction. The National Academy of Sciences defines mutation as a, "change in the sequence of one or more nucleotides in DNA. Such changes can alter the structure of proteins or the regulation of protein production. In some cases mutations result in the organism possessing these altered traits to have a greater or lesser chance of surviving and reproducing in a given environment than other members of its species."

When it comes to displays of mutation, geneticists are facing two problems in supporting evolution. First, mutations tend to scramble a genetic sequence and/or leave out parts of a sequence. The second issue geneticists are facing has to do with the observation of incorrectly assembled DNA causing negative effects in organisms. Cellular senescence is one of three negative outcomes of genetic mutation which causes affected cells to go dormant. Another outcome causes the cell to self-destruct, otherwise known as apoptosis. The third and final outcome of incorrectly assembled DNA results in uncontrolled cell division which leads to cancer and various types of tumors.

IS EVIDENCE EVIDENT?

In our natural world today, we have not seen mutation as beneficial and have yet to observe mutations that do not have negative effects. What empirical data suggest this fact would be different in the distant past? Considering the empirical data geneticists are collecting today, is it logical to claim that mutation could become beneficial given millions of years? Mutation is the predicted engine of evolution and empirical data collected through observation paints a negative picture that does not increase life expectancy nor genetic complexity. Incorrectly assembled DNA (aka mutations) tends to lower expectancy and quality of life, while healthy genetic information remains unchanged in evolutionary stasis. Douglas Futuyma said that you can not "have any evolutionary change whatsoever without mutation." The modern study of genetics has compiled data showing mutation to be negative, and no empirical data has shown mutation to be positive.

GENETIC VARIATION: Genetic variation is the cause of many different traits across the human species. Much like a band playing a song and their audio engineer mixing the levels of each instrument and microphone, our DNA can turn on and turn off varying traits. For example, we humans are able to consume dairy as infants due to a switch on our DNA that is activated. If dairy is no longer a main staple of one's diet, the switch usually is deactivated in order to save energy. The information contained in our

WHAT IS EVOLUTION?

DNA which gives us the ability to digest dairy does not disappear, the switch just gets turned off and its information is no longer being accessed, causing lactose intolerance.

When it comes to a sound board, is the audio engineer able to add more sliders and inputs on the fly, or must he work with the inputs he has available? Although adding cello to a musical composition would be an excellent choice, he cannot for two reasons. First, there is not a cellist within the group of musicians. Second, the soundboard has no available inputs to add the signal that would be coming from the proposed cello. Such is the case when observing human DNA, do we have switches in the off position containing information to grow feathers? We do, however, have switches that contain information on how to grow hair and switches that invoke male pattern baldness. If the switches in our DNA that control hair production are turned on or off, one could be quite hairy, perhaps have baldness, or have little to no hair which is called alopecia. Trait variation does indeed vary, but according to genetic research and empirical data, are geneticists observing new sequences accrue? The "soundboard" of human DNA is constantly being re-mixed and new levels of certain traits are going up and down, but are geneticists observing the number of switches increase? Is genetic information being added, or just remixed due to random sampling? If there is no observable data that shows an increase in genetic complexity then we must ask; is DNA evident evidence for evolution?

VESTIGIAL REMNANTS: Mangoes are delicious because mangoes taste so good. Mangoes taste so good, because mangoes are delicious. Can this type of logic be at all useful? An argument that uses its result as the basis of initial reasoning is a circular type of reasoning and the evolutionary claim that vestigial remnants are good evidence in support of descent with modification is circular. Webster defines a vestige as, "a bodily part or organ that is small and degenerate or imperfectly developed in comparison to one more fully developed in an earlier stage of the individual, in a past generation, or in closely related forms." Charles Darwin wrote of vestigial remnants in *The Origin of Species*, stating "Organs or parts in this strange condition, bearing the plain stamp of inutility [uselessness], are extremely common, or even general, throughout nature. It would be impossible to name one of the higher animals in which some part or other is not in a rudimentary condition." Darwin uses circular reasoning when speaking of vestigial remnants, for the idea that "higher animals" contain vestiges assumes their emergence from lower animals. According to this type of logic, another way of saying vestigial remnants are good evidence for evolution is to say that the trail left from evolution is proof of evolution.

WHAT IS EVOLUTION?

Is it logical to claim that vestigial remnants support evolution?

In addition to circular reasoning, the idea of vestigial remnants has three other questionable aspects. First has to do with the idea that devolution or the reverse of evolution takes place. The basis of Darwin's work is grounded in the idea that a simple, unicellular organism gave rise to all life on Earth by increasing in complexity and diversity through successive generations. Devolution suggests that Darwin's foundational idea of organisms moving through time from a state of "lower" to "higher" complexity is not always the case. Vestigial remnants are described as certain structures and organs that have decreased in complexity, functionality, and could possibly devolve into an eventual non-existence.

Whale evolution is described by the evolutionary community as being quite complete by starting with the entry of a primitive mammal into a tidal region. Much like the hippopotamus having aquatic tendencies, this pre-whale began to forage and hunt waterborne food sources which, according to evolution and given millions of years, the front limbs became fins and the hind limbs devolved away. Does the idea of vestigial remnants suggest that evolution has the ability to strip away functionality and cause the loss of complexity? Can natural selection beautifully gift a primitive mammal quadrupedal mobility through a slow maturation of its limbs, but then reverse the process to remove them? The question that whale evolution brings up has to do with the possibility of devolution. Does the theory of evolution encompass the shedding of highly evolved adaptations, like limbs? If so, was there a time in the history of whale evolution that whales had useless remnants of hind limbs dragging in the water as they swam? This would have been a very inefficient time for the hydrodynamics of such an organism. Looking at whales today, it is obvious that no such drain of resources is left to be weighing them down. Modern whales are so highly adapted to their environment, but so are the forms of every other organism to their environment. Like a proposed early whale dragging useless hind limbs, do we see organisms that are alive today in such a stage of gross inefficiency? What life forms are struggling with gross inefficiencies that we can observe today? If evolution is a fact, and all life evolves, doesn't that mean there should be organisms in flux with their environment right now? What about other mammals that live dually between land and water, are their limbs observably transitioning through devolution? Sea turtles also live in water, breathe air, and crawl onto land to lay their eggs. Are sea turtles observably transitioning through devolution? Is it logical to infer that a transition like whales losing their hind limbs took place if a similar transition in organisms alive today can not be observed? If Darwin's "Tree of Life" has unicellular organisms arranged

WHAT IS EVOLUTION?

near the trunk and more complex, multicellular life out on the branches; how would his tree account for the shedding of complexity? The idea of vestigial remnants suggests that devolution is just as real as evolution, so given millions of years, could a group of humans, through their descendants, evolve into a group of unicellular organisms similar to amoebas?

The second aspect to the idea of vestigial remnants brings question to the data surrounding DNA and other organisms that appear to be devolving, like flightless birds. Are geneticists observing the loss of genetic information among organisms that have "lost" original functionality of a highly adapted trait? For example, are flightless birds slowly losing the genetic code necessary to build wings? If no observable loss of genetic code has been documented in flightless birds, how logical is it to assume it happened to whales in the unobservable past?

Perhaps, the opposite is true and whales in fact do have genes in their DNA to build hind limbs. The freshwater stickleback is a fish that has gone through a process of shedding its hind stickles. Unlike the whale, biologist have found the stickled ancestors of freshwater sticklebacks still alive today in the form of marine sticklebacks. Due to natural selection caused by climate change, thousands of years ago the two populations were separated. For some reason, unbeknownst to those who study the freshwater stickleback, the gene responsible for building hind stickles has been switched to the "off" position. A similar scenario has played out with humans that find themselves lactose intolerant, the gene is present, yet it is in the "off" position. Are geneticists finding code for mammalian hind limbs still in the genetic coffers of modern whales, but in the "off" position?

Thirdly, in most natural history museums that have a whale skeleton on display, there is an inclusion of two bones hanging from where the hips of a landborn mammal would be. These two hanging bones are identified as vestigial whale hips and are the supposed remnants of when whales walked on land. What do these two bones tell us about whale evolution? Do these bones qualify as being vestigial and in a state of uselessness? According to a study published in *Evolution, the International Journal of Organic Evolution* by Matthew Dean of the University of Southern California and Jim Dines of the Natural History Museum of Los Angeles County, whale hips have shown to not be a useless vestigial remnant. Vestigial whale hips have been found to be the support structure for a multitude of tendons and muscles that control the functionality of genitalia. In fact, the size of these bones in male porpoises, whales, and dolphins is directly related to the size of male reproductive organs found in all types of cetaceans. What is being referred to as vestigial remnants of hips are actually pelvic bones that are crucial when it comes to reproduction. Research data has shown that whales with larger testes also have larger pelvic bones and those with smaller testes also have smaller pelvic bones. When speaking of the four year study conducted with Jim Dines, Matthew Dean said their "research really changes the way we think about the evolution of whale pelvic bones in particular, but more generally about structures we call vestigial... As a parallel, we are now learning that our appendix is actually quite important in several immune processes, not a functionally useless structure."

Is it logical to call the pelvic bones of cetaceans vestigial and inutile? Based on empirical data and what has been revealed about seemingly useless structures, are vestigial remnants evident evidence for evolution?

INCREASED GENETIC COMPLEXITY: Since DNA exists, it must have had a beginning and, according to evolution, at its beginning DNA must have been far less complex than it is today. From the first appearance of genetic material all the way to the complex genetic material found in humans, we should be able to observe an accruance of genetic material, an increase of complexity. In science observation is "king," so what can we see happening in human DNA that supports this idea? Many in the field of genetics would agree that our genome contains, roughly, 3.2 billion base pairs. Evolutionary biologists would agree that life began more than 3.5 billion years ago and the DNA contained inside this life was not as complex as ours. Taking this information through an equation can give us an approximate rate of increased genetic complexity. If we take our 3.2 billion base pairs divided by the 3.5 billion years it allegedly took life to become human, the annual increase of complexity in the human genome is figured to be 0.9142857142 base pairs. Evolution predicts that, over time, all life on Earth descended from a simple organism, and if true, we should be able to verify this hypothesis by empirical data.

The ever so famous "boy king" Tutankhamen is believed to have died in the year 1324 B.C.E., which is 3,338 years before the writing of this volume. If DNA supports the idea of evolution, then King Tutankhamen's genome should have approximately 3,051 less base pairs than ours today. Not surprising, the "boy king" and a few of his relatives' genomes have been sequenced. The researchers who trail-blazed this effort ascertained many intriguing details involving disease, royal incest, and a bizarre twist when it comes to ethnicity. Precision was their goal. King Tutankhamen's genome was broken down into groups as small as 100 base pairs and during the process, no one made mention of a missing 3,051 pairs. As far as anyone involved in the project could tell, King Tut was fully human and his DNA was identical to the DNA of modern humans. If humans evolved from basic self-arranging molecules, there must have been an acquisition of genetic information in the human genome. If there is no observable acquisition of genetic information between King Tut and humans alive today, how scientific is it to infer that in the distant past the opposite would have been true?

In 1991, a well preserved corpse was found in the Ötztal Alps of which researchers quickly gave the nickname "Ötzi the Iceman." It has been estimated that he lived during the Copper age approximately 5,300 years ago. Many tests have been run on this fascinating individual, but in 2012, geneticists sequenced his genome. In their research, they discovered that he was fully human and has many living descendants sharing his DNA whom reside in Corsica and Sardinia. They also found evidence that he suffered from Lyme disease. Much like the genome sequencing of King Tutankhamun, "Ötzi the Iceman's" DNA received much attention to detail. Continuing with the evolutionary inference that humans evolved from basic self-arranging molecules, there must have been an acquisition of genetic information in the human genome. Using the age of "Ötzi the Iceman" (approximately 5,300 years old) multiplied by the previously estimated annual rate of base pair acquisition (0.9142857142), if the idea of increasing genetic complexity were true, the team of geneticists should have documented approximately 4,845 fewer base pairs than modern human DNA. Yet like King Tutankhamun, "Ötzi the Iceman" has shown no observable acquisition of genetic information between him and humans alive today. How scientific is it to infer that in the distant past the opposite of empirical data would have been true?

The conundrum of unobservable increases in base pairs is exacerbated by the actual number of base pairs in "rudimentary" organisms. Darwin's tree of life suggests that the genetic material contained in unicellular organisms near the trunk would be simple. As one travels away from the trunk onto the limbs and then the branches, there in an idea that the genetic information contained in less rudimentary organisms would increase in complexity. What does the data show? We humans have 3.2 billion base pairs in our genome, and with our very complex and highly functional brains we have assumed ourselves to be the most evolved organisms on planet Earth. Darwin and his successors have assumed amoebas to be quite rudimentary and have placed them at the bottom of the tree of life, yet the Polychaos dubium (aka Amoeba dubia) has a genome with 670 billion base pairs, which is two hundred times larger than that of humans. The Asian flower Paris japonica has 150 billion base pairs, yet the highly adapted and mobile Tetraodon nigroviridis (pufferfish) only contains 390 million base pairs in its genome. Two living fossils that have been

WHAT IS EVOLUTION?

in evolutionary stasis for eons and have been assumed far more rudimentary than we humans are the Protopterus aethiopicus (lungfish) with 130 billion base pairs and the Bufo bufo (common toad) which has 6.9 billion base pairs in its genome. Countless discrepancies can be found with Darwin's tree of life when observing the reality of what the genome of organisms reveal. What empirical data supports evolution in light of varying levels of genetic complexity between "lower" and "higher" life forms?

JUNK DNA: When trying to find logical comparisons between the evolution of rudimentary organisms into complex organisms and the assumed increase in genetic complexity, geneticists are finding nothing but confusion and frustration. Many in the field of genetics have thrown their hands in the air and have claimed "Junk DNA" to account for large and illogically sized genomes. In 2001, the Human Genome Project international consortium announced that, "chromosomes have crowded urban centers with many genes in close proximity to one another and also vast expanses of unpopulated desert where only non-coding 'junk' DNA can be found." Imagine if the National Oceanic and Atmospheric Administration stated that the ocean has crowded reefs teeming with life and also vast expanses of unpopulated water where only non-essential 'junk' ocean can be found. Our understanding of the ocean is very limited with, according to the National Oceanic and Atmospheric Administration, only five percent being explored. Our understanding of genetics is also far from complete, how is the idea of "Junk DNA" any different from the unscientific idea of "Junk ocean" being a possibility? Is the blanketing of the infantile field of genetics with the general idea of "Junk DNA" consistent with true scientific inquiry?

In addition to the the idea of "Junk DNA" being in direct opposition to true scientific inquiry, it is void of empirical methodology for the following reasons. As previously mentioned, the human genome has failed to reveal an acquisition of base pairs and fails to offer an observable increase in genetic complexity. So, what about the Polychaos dubium (aka Amoeba dubia) with its 670 billion base pairs? Is it possible that its genome has been successful in revealing an acquisition of base pairs and offers an observable increase in genetic complexity? Using the same equation in regard to producing the human genome's annual rate of base pair acquisition. We can take said amoeba's 670 billion base pairs divided by the estimated 3.5 billion years life has been in existence, and we get an approximate base pair acquisition rate of 191.428571 per year. Now that an annual rate of acquisition has been figured, is this acquisition observable? Does the amazingly large genome of this interesting amoeba acquire over 190 base pairs each year? What about other strange organisms with shockingly enormous genomes, are they too acquiring massive numbers of base pairs each year? If not, when did this acquisition take place? Even if these organisms have a massive amount of useless genetic information, can we observe how and when this "Junk DNA" accrued? If a genome of a certain size exists and evolution is true, there must have been an accruance of base pairs regardless of the labeling of "junk" or not. What observable, empirical data shows that genomes have increased or decreased in size?

Another reason why the idea of "Junk DNA" is void of empirical methodology has to do with its striking similarity to the idea of vestigial remnants. Scientific research has overturned the false idea of organs like the human appendix being useless or inutile. Much like the false claims of vestigial remnants to be without function, scientific research is showing "Junk DNA" to be quite useful and far from being deemed as "junk." In 2003, a public research consortium was founded in part by the National Institute of Health and the National Human Genome Research Institute called the Encyclopedia of DNA Elements, or ENCODE for short. The goal of this organization has been to look at the vast ocean of "Junk DNA" as a new frontier waiting to be explored. Once ENCODE removed the blanket of uselessness from this treasure trove of information, research began and the consortium has been producing amazing results. The idea that a majority of DNA is junk came from the fact that only 1.22% of the information housed in our genome is used in coding strands of protein. This is because the "how" of protein strand production has been a mystery until recently. The Encyclopedia of DNA Elements has released findings that show 80.4% of "the human genome participates in at least one biochemical RNA and/or chromatin associated event in at least one cell type." This means that our current understanding of "Junk DNA" shows that a large majority of what was thought to be junk is necessary for regulatory functions. Of course there is still so much to learn and discover, but that is the point. Science is not about throwing away empirical data that causes us to second guess hypotheses. Science is about diving head first into an ocean of data that seemingly suggests the opposite of one's initial theory. In regard to DNA, is not the pursuit of truth far more satisfying than dismissing parts of a genome as "junk" in order to hold onto inferred ideas that may or may not be true?

OUR CLOSEST RELATIVE: Evolutionary biologists have been communicating to the public for quite some time now that chimpanzees are humans' closest evolutionary relative, with 95% of our DNA being the same. The claim is that in the distant past, we shared a common ancestor, but what does the data show? Unfortunately, geneticists have been unable to find viably sequenceable samples of DNA from a pre-human and/or pre-chimpanzee ancestor. The only comparisons can be made are from reading the data of sequenced chimpanzee and human genomes.

Before any comparisons can be made, one must first understand what information is contained in a genome. The first and most noticeable structures are chromosomes. These structures categorize and house the varying number of strands of DNA. In sexually reproductive organisms, the sex genes are housed in their own chromosome(s). The number of chromosomes an organism has is very important in determining what traits it will have. A second notable aspect of an organism's genome has to do with coding and non-coding sequences. Proteins and how they function are at the heart of the information contained in DNA. So, when a sequence is referred to as a coding sequence, geneticists are referring to its ability to create functional proteins. Non-coding sequences are highly functional in a regulatory capacity. A third highly discussed aspect of DNA has to do with the computer-like information stored in base pairs.

Is the data offered in regard to organisms' relationships and common ancestry straightforward? Humans' possession of 46 chromosomes and chimpanzees' possession of 48 may seem to be a trivial discrepancy, yet the detailed information in chromosomes is so drastic that many organisms with the same number of chromosomes are of extremely different species. The coyote and the chicken both possess 78 chromosomes, are relational comparisons being made because of this? Extreme difference in species is reflected with organisms that share the same number of chromosomes with chimpanzees. Although no evolutionary biologist is claiming that deer mice and the tobacco plant are close relatives, they do share the same number of 48 chromosomes with the chimpanzee. The information housed in the varying number of chromosomes an organism has is so diverse, should we not question relational ties between humans and chimpanzees with a two chromosome discrepancy? What data revealed by the quantity of an organism's chromosomes is evident evidence of humans' common ancestry with chimpanzees?

As previously mentioned, the second aspect of DNA that must be expounded upon is within the strands of coding and non-coding sequences. This delineation is very much needed due to the importance of sequences that manufacture functional proteins and the sequences that regulate the manufacturing of said proteins. This relationship is still much of a mystery, but could be analogically explained by similarities with an ink jet printer. The printhead produces the image, yet must be guided and controlled by the computer in order to produce the desired image. Like a printer, coding sequences produce the proteins necessary for life to function, yet the non-coding sequences are like a computer that regulate the process of

WHAT IS EVOLUTION?

their manufacture. When drawing relational comparisons between organisms, does the number of coding and non-coding genes seem to be in favor of inferring relational ties? For example, below is a list of some organisms and their gene types.

ORGANISM	CODING	NON-CODING
Chicken	15,508	1,558
Chimpanzee	18,759	8,681
Cat	19,493	1,855
Dog	19,856	3,774
Cod	20,095	1,541
Human	20,296	25,173
Armadillo	22,711	5,984

What data revealed by an organism's number of coding and non-coding sequences is evident evidence for humans' common ancestry with chimpanzees?

Each parent, in sexually reproducing organisms, passes along strands of information stored on the ever so famous double helix structure of DNA within its ladder-like rungs called polymers. New polymers are constructed by using half of each parent's polymer, called a monomer. When two parental monomers are joined together, a new polymer or rung on the double helix ladder is created. This pairing is the basis of genetic code and these polymers are referred to as base pairs. Much like the binary language of computers that use the two variants of ones and zeros, genomes use five variants between two varying methods. DNA strands are sequenced using the letter representations GACT, while

IS EVIDENCE EVIDENT?

RNA strands are sequenced using the letter representations GACU. The studying of these sequences is where evolutionary biologists have pieced similarities together. In the binary language of computers, green is represented by 1-1-0. If one were to convert a photograph of a human and a chimpanzee side by side in a jungle into the binary language of computers, one would instantly see similarities. The binary code 1-1-0 for the color green would be quite prevalent, yet so would the code for black, which is 0-0-0. The code for green could accurately be describing organisms that produce chlorophyll, but no species of plant could be determined. Unfortunately, all non-plant, green organisms like leaf bugs and iguanas would also fall into the same descriptor of binary code 1-1-0. Continuing to view our jungle photo in binary code, black would also be dominantly repetitive, but what would it's code represent? Perhaps both chimpanzee and human pupils would be linked to having a strong correlation, but binary code 0-0-0 would also be representational of lifeless shadows. As in the analogy of a photo converted to binary language, finding similar genetic sequences in various genomes of varying organisms also demonstrates to be inconsequential.

As previously covered in the section of this chapter labeled HOMOLOGY, the work of Sir Gavin De Beer revealed that homologous structures among differing species do not develop from the same embryonic location. Which in turn

WHAT IS EVOLUTION?

demonstrates no ability to draw genetic relationships between two species with homologous sequences because their development is not controlled by their homologous DNA. Due to this fact, what significance is there in finding similar genetic sequences in the genome of varying species?

Another interesting aspect of base pairs represented in human and chimpanzee genomes has to do with numbers. The claim is that a 95% relational tie can be made due to the study of the genetic material in our DNA, yet what percentage does just the base pair information give us? Chimpanzees have 2.9 billion base pairs in their genome and we humans have 3.2 billion. For the sake of this argument, let's say that the first 2.9 billion base pairs in our genome are a perfect match to that of chimpanzees. Without a single variant on any given polymer, there would still be 300 million differences between the two genomes. Humans have an extra 300 million base pairs. So, what would be our percentage of being genetically related to chimpanzees based solely on the first 2.9 billion polymers being a perfect match? 2.9 billion divided by 3.2 billion is equal to 0.90625, meaning that in this fictitiously error free scenario of comparing base pairs, we are approximately 90.6% related to chimpanzees. Adding in the potential mismatches, the thousands of differences in coding and non-coding sequences, and the fact that humans and chimpanzees have a different number of chromosomes, how

IS EVIDENCE EVIDENT?

scientific is it to claim a relational tie between these two organisms at all?

What genetic data supports relational ties between species? What genetic data supports the idea of descent with modification from a common ancestor? What data supports the idea that complex DNA evolved from less complex DNA? What aspect of the field of genetics is evident evidence for evolution?

Chapter 4.

WHAT EMPIRICAL DATA SUPPORTS EVOLUTION?

"All knowledge that is not the genuine product of observation, or of the consequence of observation, is in fact utterly without foundation, and truly an illusion."
Jean-Baptiste Lamarck

Science is a word we like to use frequently. There is reassurance that comes from knowing, and we believe science can bring us the answers to why we get sick, how many calories are in a brownie, what driving pattern will increase fuel efficiency, and so on. Trust is at the heart of it all. We trust science. For this reason advertisers put men and women in lab coats on their billboards, magazine ads, and television commercials. Science sells and, in fact, we don't want to buy without it. When the science behind a product is spelled out we can see how said product might fulfill a need we have. Our experience with the product may or may not be reflective of the claims that convinced our purchasing of the item, but we end up finding that out by using and testing the product ourselves. No matter how much science went into the development of a product, we engage science at home when we put our newly purchased products to the test. When we test, observe, replicate, and verify the advertising claims by using a new product; we are engaging in the empirical process of science. A scientist must be methodical in the way he or she looks at the tangible world and seeks to find an answer to their question that is firmly grounded in what can be observed, tested, replicated, and verified. In doing so, knowledge is acquired, breakthroughs are realized, and enhancement of life comes through its application. Louis Pasteur is a perfect example of this kind of scientist for he realized scientific breakthrough as a result of experimentation and now his work is improving billions of lives through pasteurization.

DEFINING SCIENCE: According to the National Academy of Sciences, science is defined as the "use of evidence to construct testable explanations and predictions of natural phenomena, as well as the knowledge generated through this process." This "process" is commonly identified as the Scientific Method of which Webster defines as, "principles and procedures for the systematic pursuit of knowledge involving the recognition and formulation of a problem, the collection of data through <u>observation</u> and <u>experiment</u>, and the formulation and testing of hypotheses." If science is knowledge acquired through observation and experimentation, then the National Academy of Sciences stating evolution to be a fact must be supported by <u>observation</u> and <u>experimentation</u>. What data has been collected through observation and experimentation that verifies speciation to be a fact? What data has been collected through observation and experimentation that displays beneficial mutation to be a fact? What data has been collected through observation and experimentation that shows the accruing of genetic information to be a fact? In regard to explanations of the natural world, what evolutionary tests have been made in the presence of observed phenomena? What are the observed, causal elements of the emergence of new species and what experiments have been developed to replicate said process? How does evolution cause an increase in biodiversity?

The "how," of evolution was in question during Darwin's time and is still debated among evolutionary biologists today. Is it logical to claim Neo-Darwinian ideas reflect empirical methodology if the "how" of said phenomena remains unable to be observed? If a scientific idea is based on phenomena that are unable to be observed, what does that say about the scientific validity of the idea? Webster describes "an order of existence beyond the visible observable universe" as being the definition of supernatural. Is there room in "science" for the inclusion of the supernatural realm? Molecular biologist and University of California Los Angeles Professor Emeritus of Biochemistry Richard Dickerson spoke of science and the supernatural when he developed his number one rule which he called the "Game of Science." Dickerson stated that science is like a game and the only rule is to "see how far and to what extent we can explain the behavior of the physical and material universe in terms of purely physical and material causes, without invoking the supernatural." The supernatural realm is that which cannot be observed, tested, replicated, and/or verified. Knowing that speciation is at the core of evolutionary theory, what aspects of evolution are within the boundaries of Dickerson's one rule? What core processes and mechanisms of producing diverse, genetically isolated species are natural and truly supported by empirical data?

SCIENTIFIC PREDICTION: Neil Shubin, PhD, is a professor of Organismal Biology and Anatomy at the University of Chicago. He is also an author, speaker, and the discoverer of the Tiktaalik fossil. In 2004, Shubin and his team set out to find a transitional fossil that would show life making the leap from water to land. Shubin predicted that if he and his team dug through rocks of "just the right age" in the Canadian Arctic, they would find the first organisms to have evolved out of their watery home. The effort was proving to be in vain, but moments before the team lost funding, the layers of eons past separated to reveal a fossil. The real question is, a fossil of what? The skull proved to be in very good condition and showed to be instantly recognizable as belonging to the crocodilian family. Unfortunately, the body of Tiktaalik was fragmented and missing a majority of its bones. Homologous comparisons are the only means of attempting to identify Tiktaalik, for the field of genetics is unable to help due to the degradation of DNA in this specimen.

In light of the poor condition of the Tiktaalik fossil, how scientific is the confirmation of Neil Shubin's prediction of finding a transitional link between humans and our waterborne ancestors? How has Tiktaalik been empirically authenticated by the evolutionary community? Has Tiktaalik confirmed evolution by verifying Shubin's scientific prediction of finding a transitional fossil in 375 million year old rocks? Has Shubin and his team made and verified a real scientific prediction? Webster defines a prediction as "a statement about what will happen or might happen in the future." The scientific method is the tried and true way of understanding the natural world and places a high value on predictions that will take place. If one were to find themselves on a quest to grasp how things work and followed the scientific method, they would commit to evidence based conjecture and hypotheses. Next, logical consequences would have to be identified which would lead to the forming of a prediction of what might happen in the future. Lastly, a controlled and replicable experiment would be carried out of which is in order to verify said prediction. Is this what Neil Shubin and his team did to verify evolution? Did Neil Shubin make a prediction about the future? Is finding a new fossil part of experimentation or is it compiling more evidence? According to science, how logical is it to say that Tiktaalik verifies Shubin's prediction?

Neil Shubin's team has predicted to find and has found evidence that leads to strong conjecture and a strong hypothesis. Based on the evidence of Tiktaalik and the hypothesis of waterborne organisms moving onto land, one should predict what the outcome of a controlled and replicable experiment would be. Shubin could have set up a controlled experiment that would have revealed one species splitting into another. Perhaps his team could have sequenced the DNA of a group of organisms, then could have run them through varying environmental pressures in order to verify an acquisition of base pairs and an increase in genetic complexity. What if Shubin had created an experiment that showed mutations to be beneficial when it comes to increasing life expectancy and biological fitness? Has Neil Shubin and his team verified scientific predictions or have they found more evidence for their hypothesis? Can we scientifically say that Tiktaalik verifies evolution? Has the discovery of Tiktaalik validated Shubin's prediction with empirical data or has it produced more evidence that is waiting to be tested?

WHAT EMPIRICAL DATA SUPPORTS EVOLUTION?

Could it be that Darwin, himself, erroneously set naturalist out on an unscientific trajectory when communicating in *The Origin of Species* that the lack of transitional fossils hurts the validity of evolution? Can new convincing evidence ever verify a prediction in a scientific sense?

OBSERVABILITY OF EVOLUTION: There is a battle raging over the validity of evolution. Some feel conflicted, yet most would agree that science has brought us breakthrough/innovation. If evolution is scientific, then the National Academy of Science's claim that evolution is a fact must be true. Louis Pasteur was a scientist and his work is not being questioned, so why is the work of Darwin and his successors? Could the tension stem from a lack of trust? Perhaps there is an overwhelming sense of uneasiness when saying "evolution is fact." Unlike bringing home a product and testing the advertising claims, we are not able to run evolution through the same process. When we trust in the work of scientists we hope their claims are firmly grounded in what can be observed, tested, replicated, and verified.

In the high profile case of Kitzmiller versus the Dover area school district, the scientific validity of intelligent design and the constitutionality of it being taught in public schools was on trial. Ultimately the court sided against intelligent design because it was found to be unobservable, untestable, unreplicatable, and unverifiable. The court also decided that intelligent design is nothing more than creationism with a scientific label attached to it. Intelligent design lost the case and will no longer have a chance of entering the public school classroom.

In the court records of this famous case, (Kitzmiller versus the Dover area school district) Professor of Ecology, Evolutionary Biology and Behavior at Michigan State University, Robert T. Pennock clearly describes that as a science-focused community we must "seek explanations in the world around us based upon what we can observe, test, replicate, and verify." In the same case, Molecular biologist and Professor of Biology at Brown University, Kenneth Miller was being cross-examined when he was asked to read Page 27 of Exhibit 649 and his reply was, "Be glad to. This is the opening of the third section of this book [Teaching About Evolution and the Nature of Science by National Academy of Sciences], and it opens basically by defining science. And it says, and I quote, Science is a particular way of knowing about the world. In science, explanations are restricted to those that can be inferred from confirmable data, the results obtained through observations and experiments that can be substantiated by other scientists. Anything that can be observed or measured is amenable to scientific investigation. Explanations that cannot be based on empirical evidence are not part of science."

WHAT EMPIRICAL DATA SUPPORTS EVOLUTION?

What if the tables were turned in this trial? What if the scientific validity of evolution were to be scrutinized? What if the cross-examiner continued to ask Kenneth Miller questions, but flipped the tables toward the scientific validity of evolution? How may have Darwin's theory stood up in a court of law? The "gold standard" of outlining evolution's scientific worth is completely intertwined with the origin of species. Can the emergence of genetically isolated organisms be observed, tested, replicated, and verified?

PROCESSES & MECHANISMS: The National Science Teachers Association is a major proponent of evolution and in their position statement the following comment is made. "There is no longer a debate among scientists about whether evolution has and is occurring. There is debate, however, about how evolution has taken place: What are the processes and mechanisms producing change, and what has happened specifically during the history of the universe?"

Does this position statement place the National Science Teachers Association in a position that values empirical data? Imagine the Wright brothers or Louis Pasteur using this kind of logic when sharing their discoveries. If Wilbur Wright were to have used the same logic as the National Science Teachers Association he might have said; Hey Orville, I am utterly convinced that powered human flight is a fact, but I am not sure how it is going to take place. What are the processes and mechanisms needed to produce powered human flight? Oh well, let's tell everyone that powered human flight is a fact and let's teach it in public schools and universities. Perhaps, if Louis Pasteur used this same type of reasoning he may have said the following while enjoying a nice picnic on a hot summer's day... My dear would you like some bottled milk with your sandwich? I am utterly sure it is safe to drink because pasteurization is a fact. I am unaware of why milk goes bad and do not know the processes or mechanisms needed to make pasteurization possible, but go ahead and drink the milk my dear.

Jerry A. Coyne perpetuates this way of thinking by writing in his book *Why Evolution is True* that there, "is no dissent among serious biologists about the major claims of evolutionary theory - only about the details of how evolution occurred, and about the relative roles of various evolutionary mechanisms." Do those who support this type of rationale truly understand science and value empirical data? When the National Science Teachers Association and prominent voices like Jerry A. Coyne share evolutionary ideas as fact yet do not detail a concise "how" or "mechanism," are they reflecting a true image of science? Observability is the lifeblood of science and if no observable speciation nor beneficial mutation can be made, is it logical to say that evolutionary thought values empirical data?

FALSIFIABILITY: Science is hugely dependent upon experimentation for it can verify a proposed hypothesis. Experimentation can also verify subsequent predictions and reveal if a proposed hypothesis is able to be proven wrong. Karl Popper's work is held in high regard among the scientific community and he concluded that true scientific experimentation can only take place if the experiment is designed to be falsifiable. Falsifiability is the idea that a hypothesis or a theory must be disprovable in order for it to be scientific. Popper was very clear in communicating that falsification is the validating factor of a theory's scientific credibility.   If a theory cannot be falsified then it is considered pseudoscience.

Is it possible to prove evolution to be wrong? What experiments have been created with the falsifiability of evolution in mind? Some evolutionary biologists have agreed that finding a fossil out of order could most definitely put Darwin's theory in question, yet can this be considered an experiment? Is finding a fossil the same as setting up an experiment that would produce empirical data? In other words, if one were to find rock layers containing mammals that were below organisms from the Cambrian, what empirical data would be collected and could that data give the ability of the scientists to falsify the hypothesis? Can evidence supplant experimentation? Would finding a fossil of any species in any layer of rock verify evolution, or just add more evidence to the hypothesis?

Let us push experimentation aside for just a moment and entertain the idea that evolution could be falsified by finding a mammal in the Precambrian. In doing so, we would find three reasons to scientifically question the validity of this claim. Firstly, Darwin had issues lining up his ideas with the reality of how certain rock strata came to be. Instead of changing his ideas, Darwin claimed there to be an "imperfection in the geological record." He stated in *The Origin of Species* that "geological formations of each region are almost invariably intermittent; that is, have not followed each other in close sequence." Geologists are still wrestling with this conundrum today. Sediments become overturned. Lava flows cut through ancient bedrock and fill their path with granite. Rivers cut layer after layer and carry fragments of older layers to new locations depositing composite layers. Tectonic forces lift lower layers and create mountain tops out of rock containing fossilized waterborne organisms. To say that finding a layer out of place would be a valid statement of evolutionary falsification is incongruous with observation, for no layer is

WHAT EMPIRICAL DATA SUPPORTS EVOLUTION?

perfectly in place. The real issue comes back to which layers are considered most ancient and which are considered to be most recent. If we are to compare fossil evidence of varying tree of life locality; timing is of the essence. So then, how might age be applied? In *The Origin of Species* Darwin gave an estimate to the erosion process when he said, "a cliff 500 feet in height, a denudation of one inch per century for the whole length would be an ample allowance," making said cliff a flat plain in 600,000 years. Yet, it is not unlikely for an entire hillside to peel off into the ocean after a heavy rainstorm. Denudation and the laying down of rock layers takes place at varying rates, and anyone's guess to the timing is up for debate. The United States Geological Survey holds the stance that, "the actual length of geologic time represented by any given layer is usually unknown or, at best, a matter of opinion." So in order to have a geologic age a bit more concrete than opinion, some believe the best way to determine the age of a certain layer is to determine what index fossils are contained in such layer. Index fossils are used to date and correlate varying rock strata. For example, if one were on a hunt to find a 540 million year old rock, then a 540 million year old index fossil would be the best clue to look for. An index fossil of the right age is what one should look for when identifying rocks from the Paleozoic, which began roughly 540 million years ago. Paleozoic strata is not comprised of one type of rock, in fact, its composition varies quite drastically and ranges from limestone to quartz-feldspar schist. One could find Paleozoic rock on a mountain top or in a riverbed. The most common determining factor when identifying this particular layer of strata is finding an index fossil.

Secondly, another reason to question the validity of a falsification statement including mammals found in the Precambrian has to do with how the age of index fossils are determined. Trilobites are what some people call a rudimentary organism and a good index fossil to look for when identifying Paleozoic rocks, but why? Look to Darwin's tree of life. It is quite obvious to see that rudimentary life is at the base of the tree and the least rudimentary is toward the top. Of course Darwin was the first to explain that natural selection was how he believed life to have become so diverse, but he was not the first to see life as changing from simple to complex. The Victorian era was ripe with young naturalists who spoke of rudimentary organisms changing into less rudimentary ones; this era was when the geologic periods were invented. Of course, the most basic of organisms were placed into the strata at the bottom of the chart and as the complexity of an organism increased, it was placed higher on the chart. Then a rough estimation of how long it might take for a rudimentary organism to diversify into more complex species was bracketed over that particular era.

WHAT EMPIRICAL DATA SUPPORTS EVOLUTION?

Thirdly, the idea of index fossils must be examined. Is claiming that fossilized organisms are contained in certain rock layers based on when said organisms evolved into existence free of bias? Does not the mentioning of index fossil presuppose the validity of evolution? So then, is it accurate to say that using index fossils to validate evolution is free of circular reasoning? How scientific is it to use the idea of index fossils when attempting to validate or falsify evolution?

According to the National Parks Service, the usual method geologists use when giving an age to a specific rock layer is by identifying index fossils. Yet, is the giving of age to rock formations based on index fossils an accurate method? Is the categorization of a fossilized organism into one of the geologic periods free of bias? If a fossilized mammal were to be found and the researcher chose to date the specimen using the index fossil method, the chemical composition of the rock encompassing the organism would not be analyzed. Mineral content and proximity to the surface would also not be a factor in determining its age. If a researcher uses the index fossil method to date a fossilized mammal, the specimen would be placed in or after the Mesozoic because it required a longer amount of time to evolve into a mammal. Yet, if a fossil of a trilobite is captured in the very same type of stone, with the very same chemical composition, containing the same trace minerals, and having a similar depth or proximity to the surface, its age would be classified as being part of the Paleozoic because it is a trilobite and according to Darwin's tree of life, trilobites are located near the base of the trunk.

Falsifiability is a grave problem for evolution and its means of applying age to a specimen further complicates this issue. Why do subscribers of Neo-Darwinian thought believe that following a chart compiled in the Victorian era and using index fossils is a good way to date fossils and rock layers? Is it logical to say that finding a mammal in Precambrian aged rock is a good way to falsify evolution?

If subscribers of evolution believe that finding a mammal in Precambrian rocks seems like a valid statement of falsification, what about finding a Precambrian aged organism alive today? Jerry A. Coyne stated that, "species of animals and plants living today weren't around in the past, but are descended from those that lived earlier." Why then do we find organisms thought to have gone extinct millions of years ago still alive today? Do these "living fossils" further exacerbate the reality that evolution is unfalsifiable? When naturalists unearth fossils containing organisms with a striking resemblance to organisms alive today, a common approach to explaining away this problem is to say that some organisms are so well adapted to their environment that they did not need to change. Does this "living fossil" explanation make sense knowing that our environment is constantly changing? Another issue with this explanation is that it is a hypothesis that predicts the opposite of the main idea of evolution. Which organisms are immune to evolution and which organisms are most susceptible? Does the adding of hypotheses that are in stark contrast to the main concept of a theory causes said theory to be unsinkable, otherwise known as unfalsifiable? If evolutionary explanations include both transitional fossils and living fossils as validation of evolution, how could Darwin's theory ever be falsified?

J. William Schopf is a renowned author, paleobiologist, discoverer of Precambrian microfossils, director of the Center for the Study of Evolution and the Origin of Life, and professor of earth sciences at the University of California in Los Angeles. Yet, he perpetuates the "unsinkable" nature of biological evolution when speaking of a 2 billion year absence of evolution in sulfur-cycling bacteria. J. William Schopf stated in an interview with ABC News that it "seems astounding that life has not evolved for more than 2 billion years — nearly half the history of the Earth." The century old *Proceedings of the National Academy of Sciences of the United States of America* published the work of J. William Schopf pertaining to said bacteria. His paper claimed the bacteria's evolutionary stasis gives promising evidence for evolution. Mr. Schopf continued by saying, "Given that evolution is a fact, this lack of evolution needs to be explained."

Essentially, J. William Schopf stated that no observable evolution is evidence of evolution because evolution is a fact. Using this logic, perhaps one might explain that no observable speciation is evidence of speciation because speciation is a fact. No observable beneficial mutation is evidence of beneficial mutation because beneficial mutation is a fact. Much like Bill Nye's explanation of "the missing nature of missing links is actually further proof of evolution," this type of rationale one could illogically verify anything.

Of course, Mr. Schopf explains the bacteria's evolutionary stasis by saying these "microorganisms are well-adapted to their simple, very stable physical and biological environment. If they were in an environment that did not change but they nevertheless evolved, that would have shown that our understanding of Darwinian evolution was seriously flawed." Does this type of logic help support the possibility of evolutionary falsification? What does this finding mean for the idea of genetic drift? Genetic drift is one of the major and well accepted hypotheses of evolution which proposes that a geographically isolated population experiences speciation due to mutation and random sampling, even in an unchanging environment. Given millions, or in this case billions of years, genetic drift would have caused enormous evolutionary change. Another issue with Mr. Schopf's logic has to do with the inverse. How might one explain observable changes in the environment, but observe evolutionary stasis? Over the eons, oceans and continents have changed in shape and size, plus the climate has drastically changed. We can observe a multitude of organisms alive today that have thrived in active and changing environments, yet have remained in stasis. Does evolutionary thought seem logical if both static and changing environments are home to organisms in evolutionary stasis? Does evolution seem falsifiable if one can claim to understand evolutionary stasis in a static environment, but then also claim to understand evolutionary stasis in a changing environment? Is empirical data valued by those who claim that evolutionary stasis can exist in both changing and unchanging environments?

Many in the field of evolution take cues from Jerry A. Coyne when he stated in his book *Why Evolution is True* that "in every case, we can find at least a feasible Darwinian explanation," yet later when speaking of explanations contrary to evolution he said, "if you can't think of an observation that could disprove a theory, that theory simply isn't scientific." Can Jerry A. Coyne and others in his field have their cake and eat it too? Are they scientific when stating all things support evolution, but then say that something is not true science if an observation disproving said theory cannot be made? Can we believe evolutionary biologists like Professor Schopf when they use an organism's evolutionary stasis when in a stable and unchanged environment to support evolution, but then also believe that an organism's evolutionary stasis in a changing and dynamic environment also supports the theory? David Deutsch is an author and Oxford Professor in the Department of Atomic and Laser Physics at the Centre for Quantum Computation. He has been very clear in communicating that, "truth consists of hard to vary assertions about reality." He goes on to say that, "easy variability is the sign of a bad explanation, because, without a functional reason to prefer one of countless variants, advocating one of them, in preference to the others, is irrational. So, for the essence of what makes the difference to enable progress, seek good explanations, the ones that can't be easily varied, while still explaining the phenomena." What does the absence of falsification, by means of its "easy variability," communicate about the scientific validity of evolution?

INFERENCE: In 1859, Darwin completed his most famous work, *The Origin of Species*, in which he proposed a hypothesis. Throughout his journeys and presumable sleepless nights, Darwin's thoughts were churning, pondering, and combining natural occurrences into one cohesive idea. He was faced with a realization that "more individuals of each species are born than can possibly survive; and as, consequently, there is a frequently recurring struggle for existence, it follows that any being, if it vary however slightly in any manner profitable to itself, under the complex and sometimes varying conditions of life, will have a better chance of surviving, and thus be naturally selected. From the strong principle of inheritance, any selected variety will tend to propagate its new and modified form." These truths of the natural world have been claimed by Darwin and are now being called "evolutionary tenets." These truths have never been in question, because they speak of observable and replicable data. What is in question is the extrapolation of these truths into something that is not observable or replicable, which is the emergence of diverse genetically isolated organisms, otherwise known as speciation. Many evolutionary biologists today use a bait and switch technique when the conjecture arrives to this point. Jerry A. Coyne wrote a book detailing why evolution is true and based its validity on the fact that "the major tenets of Darwinism have been verified." Truths of the natural world have been verified. Yes, species do produce offspring at a higher rate than their rate of survival. Yes, nature poses many struggles for existence and organisms best suited will survive. Yes, organisms that do survive will pass heritable traits to their offspring. Truths of the natural world are true, they have been verified, and give a better understanding of natural selection. Darwin did not invent the idea of evolution, he used the truths of the natural world to explain his idea of natural selection. Natural selection is true. Adaptations do take place because of natural pressures, but do adaptations eventually churn out diverse species? The truths of the natural world, otherwise called the tenets of Darwinism, explain the phenomena of natural selection. Evolutionary biologists extrapolate these truths to <u>infer</u> the emergence of diverse, genetically isolated organisms. This is the heart of evolutionary contention. This is where some erroneously stop the scientific process and become satisfied with inference.

Webster defines inference as "the act or process of <u>reaching a conclusion</u> about something from known facts or evidence." Science can logically call upon inference when forming a hypothesis. Once the hypothesis is made, experimentation is necessary to confirm the hypothesis, and inference is no longer part of the process. Charles Darwin inferred speciation from the previously mentioned tenets, but instead of setting up experiments to observe speciation, evolutionary biologists have unscientifically used such tenets as confirmation of Darwin's inference. This circular type of thinking is the reason why evolution is in question. Richard Dawkins clearly outlines in his book *The Greatest Show on Earth* that he "will take inference seriously - not mere inference but proper scientific inference - and I shall show the irrefragable power of the inference that evolution is fact." Is Mr. Dawkins correct? Can inference be used to show that something is a fact? Can the National Academy of Sciences call evolution a fact if its main idea of speciation is upheld by inference?

ASKING FOR A HOAX: Theory-ladenness has plagued evolutionary research since Darwin first released his most famous work. In 1859, he wrote of the future and how naturalists were bound to find transitional fossils which had frustratingly eluded naturalists up to that point. With little surprise, an army of fossil hunters began to comb the globe on a quest to find transitional links. Only two short years after Darwin's prediction of innumerable intermediate forms, the famous reptile-like bird, archaeopteryx was unearthed. Could the discovery of such a wonderful fossil be considered luck? Why have so few transitional fossils been found, yet why have so many fossils of organisms in evolutionary stasis been found? Was Darwin right, are transitional fossils good evidence for evolution? Why were there no transitional fossils found <u>before</u> Darwin's decree of needing them to support his theory? Thousands of fossils were available to study before the release of *The Origin of Species* and many naturalists were already engaged in the conjecture around evolution. Why then were no transitional fossils documented before 1859? Could there be example

WHAT IS EVOLUTION?

after example of transitional fossils in the coffers of Victorian aged collections? Perhaps the way to find a transitional fossil that is absolutely free of the possibility of being a hoax would be to rummage through Victorian aged collections that predate the release of Darwin's book. Why have we seen fossils with credible pedigrees later to be revealed as nothing more than the work of a forger? There is an enormous need for trust when it comes to believing the claims of those who find transitional links and that trust has been broken. Darwin's prediction of transitional links has produced hoax after hoax, yet why is the acceptance of the newest hoax so common?

No other fossils have experienced the number of forgeries as those representing bird-like reptiles. Oliver Knevitt, a writer for Science 2.0, communicated his frustration when it comes to manufactured fossils.

"Let's be clear: testing the authenticity of a fossil is incredibly difficult, as hoaxers have got incredibly adept at creating convincing fakes. Never mind that Archaeraptor (fossil of a bird-like reptile), the most prominent faked fossil of recent times, was 88 separate slabs glued together with builder's grout. This apparent bodging could only be revealed with a high resolution CT scan, not normally available to most researchers."

Why is evolution reliant upon hoax prone evidence and not actual observation, testability, replicability, and verification? If naturalists are hoping to find a missing link, is it logical to say that evolution values empirical data?

WHAT EMPIRICAL DATA SUPPORTS EVOLUTION?

QUESTIONING EVOLUTION: In his book, *The Blind Watchmaker*, Richard Dawkins stated that no "serious biologist doubts the fact that evolution has happened." The National Center for Science Education makes a similar statement by saying scientists "do not doubt that species have changed over time, that is, descended with modifications from previous species, and that all known organisms share common ancestry." Is it acceptable for one to say that no "serious biologist doubts the fact that evolution has happened?" Why must this be said? Building a scientific theory on unobservable inferences is similar to building a structure on quicksand. If one could rally enough intelligent engineers to work on the project, others would stop doubting and start believing in the possibility of building on quicksand. Would the cry of this futile endeavor eventually be; No serious engineer doubts their ability to build on quicksand? As more astutely exceptional folks join the group, others would begin to give validation to the endeavor due to association. The work of author, psychologist, and professor emeritus at UC Berkeley, Irving L. Janis produced what is called groupthink. It is best described as, "a psychological phenomenon that occurs within a group of people, in which the desire for harmony or conformity in the group results in an irrational or dysfunctional decision-making outcome."

WHAT IS EVOLUTION?

Why must the evolutionary community be so convinced of its own infallibility that if any doubt is communicated, one might find their credentials put into question by the group? Such is the case with statements like, "no serious biologist doubts the fact that evolution has happened." This ultimately validates the thoughts of prominent journalist and staff writer for *The New Yorker*, James Surowiecki when he said, "The important thing about groupthink is that it works not so much by censoring dissent as by making dissent seem somehow improbable." This statement begs for a question. Do evolutionary biologists digest valid criticism gathered from observable evidence in order to alter their stance, or do they discount empirical data and hold fast to the main beliefs of the group regardless of the scientific data? Can we honestly say there is scientific value to groupthink? Richard P. Feynman said, "Religion is a culture of faith; science is a culture of doubt." If no serious biologist doubts evolution, what does that say about the culture of serious biologists?

Chapter 5.

IS FAITH NECESSARY?

"We need to define religion as 'belief in something supernatural'."
Richard Dawkins

The American way of doing business is a product of competitive innovation. A battle is raging within the boundaries of our consumptive society. Each service, product, and ideal finds itself clamoring for market share in direct competition with like entities. For generations, Coca-Cola has been engaged in a full blown war with Pepsi. Which cola is America's favorite? Well, it's still up for debate and the same is true in the battles being fought between Ford and Toyota, iPhone and Android, & Nestle and Hershey. Which car company is acquiring the greatest share of the market? Which smartphone company is acquiring the greatest share of the market? Which chocolate company is acquiring the greatest share of the market? This type of head-to-head competition continues. Which publisher, movie studio, record label, shoe manufacturer, fast food chain is holding the highest level of market share? Companies engage the consumer market knowing there is competition, yet not all entities are in

WHAT IS EVOLUTION?

competition with each other. Android is competing with other smartphone companies and not in competition with Coca-Cola. Similarly, McDonald's is competing with other fast food restaurants and not Toyota.

HEALTHY COMPETITION: When the time comes to choose what we think will best serve our needs, companies hope the fighting for market share will pay off and they will be the dominant brand in our minds. The battlefield encompassing how we got here and how it is to be taught to the next generation, is no different than companies engaged in healthy competition for market share. Evolution has one major competitor and it is made up of those who believe in the supernatural causation of biodiversity. The overarching presence of a creative, supernatural force bringing life to this humble planet is religion and it is in direct competition with evolution. Many things can be learned from each competitor in the battle over market share and the first way to do so is to look at the competitor on the other side.

Healthy competition comes between like entities; if a company finds itself in battle with a soda company, it is safe to say it is also a soda company. If a company finds itself battling a shoe company, it is safe to say it is also a shoe company. Philip Kitcher received his PhD from Princeton University and is currently a John Dewey Professor of Philosophy at Columbia University and a renowned author.

IS FAITH NECESSARY?

Kitcher not only agrees that evolution is at odds with religion, but in his book, *Living with Darwin*, he specifies the type as being a "supernaturalist and providentialist religion." Evolution's biggest competition is nothing other than the idea of a creative and supernatural being. If evolution is engaged in a battle over market share with a certain type of religion, what does that tell us about evolution?

THE MAGIC OF MILLIONS: Unusual phenomena surround the central themes of most global religions, and ancient texts describe how basic laws of physics tend to get broken. Supernatural events are described as have taken place and those who bear witness found themselves "relating to an order of existence beyond the visible observable universe," which is Webster's definition of supernatural. Of course, there is no way to observe, test, replicate, nor verify similar phenomena today. Events that are "beyond the visible observable universe" are not able to be verified and their acceptance can only be through faith in the credibility of recounting witnesses of such an occurrence.

Ancient texts of major religions claim their supernatural events to have taken place in the presence of a host of witnesses. Although fantastical events from religious books are beyond the observable universe, there is a claim that at one point some eyes have seen, some ears have heard, and some hands have felt. In fact, ancient texts go to great lengths to include many details of the who, what, and where in order to build observational credibility. Yet, Richard Dawkins wrote in *The Greatest Show on Earth* that "the vast majority of evolutionary change is invisible to direct eye-witness observation." So then, what evolutionary change is visible to direct eye-witness observation? Is it possible to observe speciation? If not, does that mean speciation relates "to an order of existence beyond the visible observable universe?" Does the inability to empirically observe speciation make it seem less like science and more like a supernatural phenomena? Why do people subscribe to evolution if its main idea of the emergence of diverse, genetically isolated organisms requires faith in supernatural phenomena?

Every religion has its source of power and Neo-Darwinism has millions upon millions of magic millennia. For when a faithful subscriber of evolution is in doubt, a loving "cleric" will step in and instruct the "doubters" to think of how change takes place, and given millions of years, fish with lobed fins crawl and then walk on dry land. Reassurance comes quickly with the magic of millions. When presented with a scenario unfolding over eons, why do we tend to disregard the necessity of scientific observation and empirical data? The supernatural power of evolutionary thought is the disassociation of empirical methodology in the presence of millions of years. For when a question is posed, the evolutionary silver bullet statement is "<u>given enough time</u>."

In the ancient scriptures of Judaism, there is an account of Jehovah consuming a drenched sacrifice on a water soaked altar, but no one in all of Israel is able to replicate the consuming power of Jehovah today. According to their faith it did happen in the distant past. Christians believe that Jesus walked on water, but physics demonstrates it to be impossible. Yet, followers of Christ truly believe that in the distant past the laws of physics were broken and their Lord walked on water. The subscribers of Islam believe Muhammad to have ascended to heaven from the rock under the dome, yet no subscribers of Islam ascend to heaven today. Muslims hold fast to the belief that it took place in the distant past. Those who find themselves as subscribers of evolutionary thought believe speciation to be true even though it cannot be observed. They believe that somehow, and given millions of years, conditions were different from the observable world today. Yet, if we cannot observe the core aspect of evolution, then what other than a "supernatural phenomenon" would be the explanation even if we were given millions of years? Evolutionary thought is powered by the belief in speciation that can only come through the magic of millions. Life's biodiversity through evolutionary speciation is completely dependent upon proposed occurrences that are "beyond the visible observable universe." Evolution is supernatural and according to Richard Dawkins, we "need to define religion as belief in something supernatural." Why are supernatural explanations being taught as fact in public classrooms?

No religion can verify the claims it stands for, the devout believe in religious claims solely by faith. How are subscribers of evolution different than faithful parishioners that adhere to a system of beliefs for which there is no observation? The great Charles Robert Darwin made a decree that still reverberates through the hearts and minds of an unwavering flock, "whoever is led to believe that species are mutable will do good service by conscientiously expressing his conviction; for thus only can the load of prejudice by which this subject is overwhelmed be removed." As devout followers of Neo-Darwinism, one must stay true to their conviction and do a good service of expressing and sharing their faith which is not observable, testable, replicable, nor verifiable. Yet, subscribers of evolutionary thought are polite in proselytizing, for in his book, *Why Evolution is True*, Jerry A. Coyne wrote of the improbability of converting everyone to evolution and how it might be too much "to assume that *The Origin of Species* can supplant the Bible." Is Coyne trying to reign in the hopes of the evolutionary devout and their ability to persuade those with divine conviction?

RELIGIONS THAT COEXIST: It only takes one afternoon to drive about the city and find a playfully hopeful bumper sticker that reads "COEXIST" written with letters in the shape of religious symbols. Deep down inside, most would enjoy very much for this to be a possibility. If we could all believe what we want to believe and no one would disagree, that would be great. The issue with this bumper sticker and this way of thinking is that it is false. Yes, we can coexist in a social sense, but there is real energy being expended on the spreading of ideals, ideas, and beliefs of which varying faiths will never see eye to eye. Does any healthy person desire conflict? Most would enjoy for us all to agree. Unfortunately, one does not have to read very far into any religious text to quickly understand the religion at hand is ultimately exclusive. There are huge discrepancies in the theology of varying faiths which will never be reconciled without eradicating the opposing faith. Wars are raging in an actual sense with tanks and guns, but there are also wars raging on an emotional, philosophical, and educational front. When speaking of theistic providential supernaturalists, Richard Dawkins said, "that evolution is fundamentally hostile to religion. I've already said that many individual subscribers of evolutionary biology, like the Pope, are also religious, but I think they're deluding themselves. I believe a true understanding of Darwinism is deeply corrosive to religious faith." Jerry A. Coyne communicated similarly in a presentation at the Harvard Museum of Natural History that "if you want people to accept the fact [of evolution]... you'd have to get them to reject their faith." He and many other subscribers of evolutionary thought know that conflicting faiths will always conflict unless the opposing faith undergoes eradication. Top position holders in the field of education are on a mission to rid society of opposition to evolution and have narrowed their sites on theistic providential supernaturalism. Evolution is not just a religion with its own supernatural events, but it is a religion that is actively trying to expunge another religion much like Jihad or Holy War. When living in an age filled with talk of acceptance and bumper stickers flaunting ideas like "COEXIST-ing," it is difficult to believe that we have such tolerance of one faith being so honest about its plan to annihilate another faith. It seems contrary to the ideals of religious freedom these United States were founded upon. If the evolutionary community is hard at work to eradicate religious opposition to their ideas, how can tax dollars be justly used in evolution education?

Chapter 6.

IS SPENDING TAX DOLLARS ON EVOLUTION LEGAL?

"To compel a man to furnish funds for the propagation of ideas he disbelieves and abhors is sinful and tyrannical."
Thomas Jefferson

Taxation is a word that arouses emotion. High is the price we pay to live in a society with well maintained roads, dependable food supply, national security, and patrolled streets among other things. The amount taken from the fruit of our labor is sizable, but when it goes to something we believe makes this country better, some of us feel taxes can be justified.

The negative emotions equated with taxation subside with the justification of costly expenditures related to things that improve this great land, yet the opposite is also true. A sinking feeling can grow overwhelmingly deeper when the spending of tax dollars is on something less likely to be classified as justifiable. We all could name a great many things upon which, tax dollars could be wasted. To imagine

WHAT IS EVOLUTION?

that every hour we spend working, a real number of minutes are going to pay for something with which some do not agree. Unfortunately, there is not much we can do about this conundrum, for the politicians in power are given the ability to best represent "we the people." Their policies will sometimes be in our favor, other times will be in the favor of someone else, and such is life. Combating the "unjustifiable policies" that are not in our favor can be legally done on the battlefield of the local place of polling.

We have the legal ability to vote our opinions into the realm of justified tax dollar expenditures, but what if the line of legality is crossed? How can tax dollars ever be justly spent on something that is against the law? The sinking feeling attached to the already dwindling paycheck grows ever more unbearable when realizing our hard earned money is taxed in order to fund illegal activity. Many have taken the "such is life" approach and have continued to allow their pockets to be picked in order to support crime on a governmental level, but "we the people" must take a stand. Could it be that using tax dollars to research, popularize, and teach the theory of evolution is unconstitutional?

LEGAL PRECEDENT: Keep in mind, this volume is by no means in support of creationism nor intelligent design, yet many of the legal issues pertaining to evolution were brought to light in court cases surrounding this debate. The

IS SPENDING TAX DOLLARS ON EVOLUTION LEGAL?

case of Kitzmiller versus the Dover area school district concluded that intelligent design is not science and is unable to separate itself from religion, therefore violating the Establishment Clause located in the First Amendment. If one were to read through court transcripts of this case and insert evolution in the place of intelligent design, at a great number of places, evolution would fare just as poorly. The essence of the case revolved around the undeniable fact that science pertains only to that which can be observed, tested, replicated, and verified. It also was very clear in establishing that scientific explanations can only be made when based on empirical evidence.

Let us take a moment and actually put evolution on trial with the same legal and scientific criteria that discharged intelligent design from the public classroom. What part of evolution can be observed, tested, replicated, and verified? What empirical evidence supports speciation? Has anyone observed a new species come into existence? If empirical evidence shows species to have remained unchanged for eons, why would true scientists infer otherwise? If every available sample of genetic material has revealed speciation to not have taken place, why would true scientists infer otherwise? What aspect of evolutionary theory is layered upon explanation after explanation derived from empirical evidence, instead of through theory-laden inference? When "science" succumbs to basing its explanations on untested inferences, what happens to the value of observation and experimentation? Is there room for theory-laden and inference based explanations in science?

The 1981 court case Segraves versus State of California concluded that scientific explanations pertain to the "how" and do not pertain to an "ultimate cause." Yet, the National Science Teachers Association holds fast to the notion that there is "no longer a debate among scientists about whether evolution has and is occurring. There is debate, however, about <u>how</u> evolution has taken place: What are the processes and mechanisms producing change, and what has happened specifically during the history of the universe?" According to legal precedent, can evolution be considered a scientific explanation if it does not pertain to the "how?" Can we <u>honestly</u> <u>teach</u> "science" that does not explain the "how?" Can we <u>legally</u> <u>teach</u> evolution if it doesn't explain the "how?" Why are we spending taxpayers' money on non-science that should be going to actual science? Explaining the "how" through empirical methodology is at the core of science. Let's put claimed evolutionary "processes and mechanisms" to the test. Powered human flight was claimed to have been possible, but the Wright brothers tested and confirmed the "how." Teaching aeronautics and the journey it has gone through is actual and legal science. If evolution has no observable, testable, replicable, nor verifiable "how," does not the teaching of it seem illegal and just embarrassing to science?

Most papers dedicated to Neo-Darwinian "research" disclose a source of funding with either the National Institutes of Health and/or the National Science Foundation. In the United States, unscientific practices are funded with taxpayer's dollars. Of course there is not much we can do about other countries, but one could imagine the enormous annual waste of public funds on a global scale. In our country, let's hold a standard of science that is not only in promotion of empirical methodology, but also hold a standard that is reflective of legal precedent. Evolution is not legal science unless the "how" can be observed, tested, replicated, and verified.

EVOLUTION IS SUPERNATURAL: The core aspects of what drives evolution, otherwise known as the "how," are incapable of succumbing to scientific explanation. Subscribers of Neo-Darwinian evolution presuppose an existence of a non-random selective force which guides all life into unobservable speciation. Paradoxically, in order to fully believe in the naturalistic explanation of biodiversity, one must disassociate themselves from naturalistic observation and put their trust in something beyond the scope of empirical methodology. There is a hope deep down in the hearts of subscribers of evolutionary biology

WHAT IS EVOLUTION?

that, given millions of years, something must have taken place contrary to what is naturally observed today. If one were honest with themselves, every Neo-Darwinian thinker would admit the "how" of evolution is utterly powered by faith in supernatural events that unfold over the course of eons.

Webster defines supernatural as being part "of or relating to an order of existence beyond the visible observable universe." If science cannot observe it, test it, replicate it, nor verify it... than it is outside of the natural world and therefore supernatural. When those in evolutionary circles begin conversation in regard to species, a dependence upon supernatural phenomena becomes quite apparent. How did the myriad of diverse species come to be? That is the question we have rolling through our minds today and that was the question Darwin attempted to answer when he wrote *The Origin of Species*. The National Academy of Sciences calls the emergence of species, "macro-evolution" and has stated that this large-scale evolution occurs "over geologic time." Yet, is the relationship between new species emerging into existence and large expanses of geologic time one that can be explained naturally or is it outside of observation and therefore supernatural?

Paleo-anthropologists agree that humans first began domesticating wolves approximately 100,000 years ago. In modern times, Dalmatians and Dachshunds exist and are both descendants of the wolf. If one were to classify these three types of dogs using the same method used to compare hominid skull similarities, the result would be three unique species. Yet, all dogs including wolves, Dalmatians and Dachshunds could have viable/fertile offspring with each other and are, in fact, the same species.

When it comes to homo-sapiens, paleo-archeologists believe we emerged from a human-like ancestor approximately 1.8 million years ago. Fast forward to our day, and we humans are represented in various forms including Swedes, the pygmy tribes of Central Africa, the people of Japan, the Aboriginal tribes of Australia, people who suffer from acromegaly like Andre the Giant, and Little People who have what is called dwarfism. Our varying faces and shapes of humanity are one species and we are all able to interbreed.

Subscribers of evolution would stop and say that 100,000 years of breeder selection is not long enough for new species of dogs to emerge from a wolf. They would also claim the 1.8 million years of human existence has not been long enough for each ethnic group to become a unique species. There is an evolutionary consensus that millions and sometimes billions of years are necessary for evolution to produce a new and unique species. When it comes to the emergence of new species, does the evolutionary necessity of millions and billions of years communicate that natural observation has value? How is stating that evolution produces new species over geologic time different from saying there is a phenomena that exists "beyond the visible observable universe?"

The federally funded National Academy of Sciences has stated that new species form "over geologic time." The relationship between new species emerging into existence and large expanses of geologic time is one that can be only explained outside of observation. Teaching that new species emerge "beyond the visible observable universe" is clearly in service of the supernatural. Webster defines religion as "the service and worship of God or the supernatural." Evolution does not include "the worship of God," but it does include the "service of the supernatural." On December 15, 1791 the Establishment Clause was added as part of the first Amendment to the United States Constitution which clearly states that "Congress shall make no law respecting an establishment of religion, or prohibiting the free exercise thereof." If teaching evolution promotes dependence upon the supernatural, according to the First Amendment, should Congress respect it or prohibit it?

IS SPENDING TAX DOLLARS ON EVOLUTION LEGAL?

"Parishioners" of evolutionary thought do not recognize their religion by steeple, for their support of the supernatural and proselytization takes place in the halls of academia. Yet, somehow taxpayers have been flipping the tab. Using public funds to proselytize the belief in something that is contrary to natural observation is unconstitutional. Funding the support of supernatural explanations with tax dollars is a crime. According to the Establishment Clause located in the First Amendment, the governmental backing of evolution is unconstitutional and conflicts with legal precedent. Until the day comes when subscribers of evolutionary thought can explain what they believe to be true in the absence of the "service of the supernatural," we must say stop evolutionary funding.

ERADICATION OF OPPOSITION: This great country was founded upon the idea that freedom of religion is paramount to the human spirit. Our forefathers not only wanted to worship the way they saw fit, but also desired to be free of a government that imposes a set of beliefs and practices upon its subjects. In the First Amendment, this way of thinking was given greater detail by the Establishment Clause. The people that founded our great country valued the importance and necessity of keeping a government from supporting one religion over another. Many inhabitants of the young America were afraid that a state sanctioned set of beliefs and practices would ultima-

WHAT IS EVOLUTION?

tely bring discrimination toward those who subscribe to a different set. Coupled with the weight, power, and financial support of a governing body, a state religion could ultimately cause the eradication of other faiths.

Today, these United States are facing the quandary our forefathers had attempted to avoid. The United States government is funding an overt intent to eradicate faiths that discount, oppose, or inhibit the acceptance of evolution. Truancy laws have been set into motion, public schools are required to teach evolution, and science teachers run the risk of an ugly lawsuit if they happen to convey any doubt of evolution being an absolute fact. The youth of today are being taught Neo-Darwinian supernaturalism (evolution) on the public dime. As unsettling as that may be, those involved in producing the curriculum have an ever more unsettling agenda. The Association for Politics and the Life Sciences published a paper entitled, *Believers and disbelievers in evolution*, of which described "Christian religiosity" to be the "strongest correlate of disbelief in evolution." Experts in the field of evolutionary biology are actively attempting to make the absorption of their field of study more successful by setting their sites on religious faith.

When speaking of getting people to accept the fact of evolution, Jerry A. Coyne said, "you'd have to get them to reject their faith." Mr. Coyne goes on to explain that learning about evolution is the process of getting people to reject theism. If evolution is taught at a young enough age, Coyne says it "turns you into an atheist, and certainly for some people that's true. I think that happened to Darwin for example. And certainly people who go into evolution, eventually a lot of them, their religious belief tends to be retired." When asked about his own journey, Richard Dawkins said that what finally made him "into an atheist was the realization that there was no scientific reason to believe any sort of supernatural creator. That came with the understanding of Darwinian evolution."

The fact that evolution's full acceptance is dependent upon the rejection of faith in a creative supernatural force is not shocking nor appalling. Many other religious groups are similar in trying to eradicate faiths that oppose their core beliefs. Even among certain faiths there are subgroups that are actively attempting to eradicate opposing subgroups, take Ireland for example. The evolutionary community has every right to eradicate opposition to its beliefs. The above quotes by Coyne and Dawkins are perfectly acceptable and consistent with what it means to live in a country that values freedom of speech and freedom of religion. They represent a large number of subscribers of evolution and have become public mouthpieces in their own right. Millions of their books have been sold, millions of dollars have been spent to further their quest, yet the difference between writing a book and publishing such ideas in a scientific journal has to do with who is paying for the eradication of religious opposition to evolution. The scientific journal *Evolution, the International Journal of Organic Evolution*, which is funded by the National Research Council and the National Academy of Sciences stated that the "prevalence of religious belief in the United States suggests that outreach by scientists alone will not have a huge effect in increasing the acceptance of evolution, nor will the strategy of trying to convince the faithful that evolution is compatible with their religion. Because creationism is a symptom of religion, another strategy to promote evolution involves loosening the grip of faith on America."

If educators in the field of evolutionary biology are keen on "loosening the grip of faith on America," they have the right to do so, but not with tax dollars. Evolution is undergirded by a clear and honest agenda of the eradication of evolution-opposing faiths. Its publicly funded and required evolutionary curriculum is unconstitutional. The governmental backing of Neo-Darwinian thought with public funds is a crime. Until the day comes when evolution's goal shifts from getting people to reject evolution-opposing faith, we must stop using public funds to support evolutionary thought.

CONCLUSION

"The future depends on what we do in the present."
Mahatma Gandhi

WHAT NOW?: The future will ultimately reveal if the United States continues to innovate and discover new scientific breakthroughs. Yet, how the future potentially unfolds is up to those of us who are here and now in the present. What approach to knowledge and the discovery of how things work has brought mankind measured prosperity? If we study the innovators that have ushered in waves of technological breakthrough, what would we learn? To be prepared for the future, we have to place a high value on the science of empirical methodology while teaching successive generations to do the same.

No matter what Americans believe when it comes to the origin of life's biodiversity, most of us can be in hopeful agreement that a bright future is what we desire for this country. The United States has not only been at the forefront of many industries, but has actually realized their inception. We have been steadily opening new doors that have fostered new ideas in the minds of each generation. Any technological greatness this country has come to savor is due to the focus of time, energy, and funding we have shown each successive generation in the area of scientific progress and innovation. Dr. Michio Kaku is a renowned author, television personality, and Professor of Theoretical Physics at the City College of New York. He has clearly communicated that science is the "engine of prosperity" and the wealth of civilizations today has come from scientific discoveries. Therefore, if we are to truly embrace what we have learned from our past and continue to be a society that is driven by prosperity, we must stay ever mindful of how and with what we are shaping the minds of the generation that will follow us.

PREPARE THEM FOR PROGRESS: The question of what should be taught in the public science classroom must be posed. How do we shape the minds of each successive generation in order to prepare them for scientific progress, innovation, and breakthrough? What would the future look like if we asked each successive generation to focus on the reality of issues facing mankind today? What if real questions were asked of science students, ones that would ask <u>how</u> to address hunger, gridlock, energy sustainability, waste, medicine, water scarcity, and terrorism to name a few? Is learning how to satisfy an inquisitive itch the same as learning how to quench one's thirst with clean abundant water, offer speedy and efficient transportation, or end the pangs of hunger with smarter ways to farm? Adopting an approach of <u>teaching how science works</u> is far greater than <u>teaching how to infer conclusions</u> from a biased view of evidence. If the youth of today are taught to face problems and are asked to address such issues in the light of scientific progress and innovation, our society would be in a position of continued and sustainable breakthrough that will bring wealth and prosperity for generations to come.

FOCUS ON WHAT WE KNOW: So, where does this leave us when it comes to explaining the origin of species? Life's diversity is a fact. Life's diversity <u>is</u>, therefore most would agree that life had an origin. Yet, this topic has been plagued by contention for generations, and throughout this contention there has been an unhealthy and smug arrogance in believing there is a correct explanation of origins. Can we truthfully speak of <u>how</u> life became so diverse in any way other than by acknowledging our inability to scientifically know? Will true innovators and scientific minds of the future continue spending time and tax dollars on touting inferred conclusions to the origin of life's diversity? Will it someday be understood that, scientifically speaking, we will never know how biodiversity came into existence? If we desire scientific breakthrough, innovation, and prosperity; we must teach true scientific "gnosis," which is Greek for "knowledge." To say we have gnosis, means we know. Inversely, to not know is called "agnosis." When it comes to origins, what other approach is more logical and honest than to claim scientific agnosis? Being honest about our lack of knowledge in how diverse species originated is not something to be ashamed of. Admitting that no empirical data can be produced, furthers scientific investigation and protects us from becoming erroneously dogmatic. The only honest approach to origins is to claim empirical agnosticism, for scientifically we do not know how life became so diverse.

Nobel Prize winning physicist Richard P. Feynman offered a similarly elegant idea by saying, "I can live with doubt, and uncertainty, and not knowing. I think it's much more interesting to live not knowing than to have answers which might be wrong."

Curricula that engages the origin of life's biodiversity support a fact-through-inference approach to science. We must ask if this approach produces anything close to being deemed an "engine of prosperity." In order to maintain our scientific edge, the United States must agree to teach empirical methodology and set aside hope-filled philosophy. Do we believe our children's precious time and our tax dollars should be spent on anything other than science in our nation's science classrooms?

CONCLUSION

Currently, spores of confusion are being spread and evolutionary biology remains undefined. The National Academy of Sciences does not have a single, concise, and agreed upon definition of evolution. The Next Generation Science Standards, the National Research Council, the California Department of Education, the Los Angeles Unified School District, and the United States Judicial System have all failed to produce a legal, concise, and scientifically agreed upon definition of biological evolution. Yet, evolutionary curricula is required to be taught as fact. Textbooks in the hands of science students across this country do not have a scientifically agreed upon definition. In fact, those whom are producing the Next Generation Science Standards have clearly stated they "do not prescribe a curriculum nor any vocabulary list with definitions." Does this seem just? Does this further exacerbate the confusion surrounding biological evolution? Is it logical to require a topic that has no standard definition? If we do not have a definition of evolutionary biology, how can we logically teach it?

In addition to questioning why evolutionary biology has not been defined, what empirical data supports evolution? As covered in this volume, the fossil record is void of evolutionary support and is in oblate contrast with Neo-Darwinian ideas. Genetics and the data surrounding DNA do not offer a shred of evidential support for descent with modification and actually discount common ancestry all together. Why do so many folks have such conviction to teach science that is void of empirical methodology? The issue with teaching bad science is not only that it is embarrassing, but it is also contrary to legal precedent. In the 1981 court case Segraves versus State of California, the decision was made that scientific explanations must pertain to the "how," yet evolutionary biologists have been very honest about their disagreement of "how" evolution works. In the court records of Kitzmiller versus the Dover area school district, Robert T. Pennock clearly describes that as a science-focused community we must "seek explanations in the world around us based upon what we can observe, test, replicate, and verify" of which the result and means of biological evolution are incapable.

Darwinian evolution is not only unable to be observed, tested, replicated, nor verified (making it illegal science) but it is also unconstitutional. Believing that over millions of years something took place that is contrary to empirical data is describing "an order of existence beyond the visible observable universe," which is Webster's definition of supernatural. The service of the supernatural with tax dollars is unconstitutional and infringes upon our civil liberties that have been outlined in the Establishment Clause in the First Amendment to the Constitution of the United States of America.

CONCLUSION

In fact, evolutionary curricula is unconstitutional for more than one reason. Educators in the field of evolutionary biology have realized that steps must be taken to combat what is slowing the absorption of Neo-Darwinian ideas, and they have set their sights on faith that supports theistic creation. The scientific journal *Evolution* which is funded by tax dollars has stated that a meaningful "strategy to promote evolution involves loosening the grip of faith on America." How is it not an absolute infringement of our civil liberties when tax dollars are used to loosen the grip of faith on America?

We must work to keep the United States a nation that is scientifically competitive and innovative. We must teach true science that causes young minds to seek out ways scientific ideas might be wrong. We must refrain from teaching pseudo-science that seeks its own confirmation while incapable of producing empirical data. Actual science should be taught in our nation's classrooms and tax dollars should no longer be spent on denying us of our civil liberties.

Stand Up For the Future. Fight With Us. Get Involved.

www.ReadySetQuestion.com

REFERENCES

chapter 1.
WHAT IS EVOLUTION?

California Department of Education. Science Framework for California Public Schools. CDE. Sacramento. 2004. PDF file.

Coyne, J.A. 2009. *Why Evolution is True*. New York: Penguin Group.

Dawkins, R. 2009. *The Greatest Show on Earth: the Evidence for Evolution*. New York: Free Press.

Dawkins, R. Interview by CNN. *Darwin and the case for 'militant atheism'*. Darwin and the case for 'militant atheism', 2009. Web. 24 Nov. 2009.

Darwin, C. 1872. *The Origin of Species by Means of Natural Selection, or the Preservation of Favoured Races in the Struggle for Life*, 6th edn. London: John Murray.

"Definitions of Evolutionary Terms." Science, Evolution, and Creationism, National Academy of Sciences and Institute of Medicine. 2008. Web. August 2015

"evolution." Google.com. Google, 2014. Web. December 2014.

"evolution." Merriam-Webster.com. Merriam-Webster, 2014. Web. December 2014.

"Evolution." National Center for Science Education. 2008 Web. September 2015.

Futuyma, D.J. 1986. *Evolutionary Biology*. Sunderland: Sinauer Associates.

Moran, L. "What is Evolution?" The TalkOrigins Archive: Exploring the Creation/Evolution Controversy, January 1993. Web. December 2014.

"NSTA Position Statement: The Teaching of Evolution." National Science Teachers Association. n.p. July 2013. Web. April. 2015.

Nye, B. 2014. *Undeniable: Evolution and the Science of Creation*. New York: St. Martin's Press.

"Overview: NAS Mission." National Academy of Sciences. Web. August 2015

Rivera, M.C. and Lake, J.A. "The ring of life provides evidence for a genome fusion origin of eukaryotes." Nature September 2004, 431 (7005): 152–5 | DOI:10.1038/nature02848. Nature, 2004. PDF file.

Russell, W. et al. (2005), CONCEPTS IN BACTERIAL VIRULENCE: CONTRIBUTIONS TO MICROBIOLOGY, VOL. 12. DOI: 10.1159/isbn.978-3-318-01116-6

"Views about human evolution." Religious Landscape Study. Pew Research Center. Web. January. 2016.

chapter 2.
WHAT ARE SPORES OF CONFUSION?

Coyne, J.A. 2009. *Why Evolution is True*. New York: Penguin Group.

Darwin, C. 1872. *The Origin of Species by Means of Natural Selection, or the Preservation of Favoured Races in the Struggle for Life*, 6th edn. London: John Murray.

Dawkins, R. 2009. *The Greatest Show on Earth: the Evidence for Evolution*. New York: Free Press.

"Definitions of Evolutionary Terms." Science, Evolution, and Creationism, National Academy of Sciences and Institute of Medicine. 2008. Web. August 2015

"Happy 200th, Darwin!" Understanding Evolution. Berkeley, University of California. Web. August 2015

Mayr, E. 1988 *Toward a New Philosophy of Biology: Observations of an Evolutionist.* Cambridge: Harvard University Press.

Nye, B. 2014. *Undeniable: Evolution and the Science of Creation*. New York: St. Martin's Press.

Grant, P. R. 1999. *Ecology and Evolution of Darwin's Finches*. Princeton: Princeton University Press.

chapter 3.
IS EVIDENCE EVIDENT?

Church, G. M.; Gao, Y.; Kosuri, S. 2012. 'Next-Generation Digital Information Storage in DNA', Science, 337, 6102-1628

Coyne, J.A. 2009. *Why Evolution is True*. New York: Penguin Group.

Darwin, C. 1872. *The Origin of Species by Means of Natural Selection, or the Preservation of Favoured Races in the Struggle for Life*, 6th edn. London: John Murray

Dawkins, R. 2009. *The Greatest Show on Earth: the Evidence for Evolution.* New York: Free Press

de Beer, G. 1971. *Homology: An Unsolved Problem.* Oxford: Oxford University Press

Dennett, C. 1995. *Darwin's Dangerous Idea: Evolution and the Meaning of Life.* New York: Touchstone

Dieckmann, U. 2004. *Adaptive Speciation.* Cambridge: Cambridge University Press

Dunham, I. and Kundaje, A. et al. "An Integrated Encyclopedia of DNA Elements in the Human Genome." Nature September 2012, 489(7414): 57-74. | doi: 10.1038/nature11247. Nature, 2012. Web.

Foote, M.; Miller, A. I.; Raup, D. M.; Stanley, S. M. 2007. *Principles of Paleontology,* 3rd edn. New York: Macmillan

Futuyma, D. "Natural Selection: How Evolution Works." American Institute of Biological Sciences, December 2004. Web. December 2014.

Gates, B. 1995. *The Road Ahead.* New York: Viking Books

Gauch, H. G. and Gauch, H. G. Jr. 2012. *Scientific Method in Brief.* Cambridge: Cambridge University Press

Goldstein, E. 2014. *Cognitive Psychology: Connecting Mind, Research and Everyday Experience.* Boston: Cengage Learning

Goodman, N. G. 1931. *The Ingenious Dr. Franklin: Selected Scientific Letters of Benjamin Franklin.* Philadelphia: University of Pennsylvania Press

Gould, S. J. 2009. *Punctuated Equilibrium.* Cambridge: Harvard University Press

Griffiths, S. "Whale sex revealed: 'Useless' hips bones are crucial to reproduction - and size really matters, study finds." Daily Mail, Science & Tech. Associated Newspapers Ltd. Web. 11 September 2014.

Hawass, Z. et al. 2010. 'Ancestry and Pathology in King Tutankhamun's Family', The Journal of the American Medical Association, 303, 638-647

Howard Hughes Medical Institute. "The Making of the Fittest: Evolving Switches, Evolving Bodies MODELING THE REGULATORY SWITCHES OF THE PITX1 GENE IN STICKLEBACK FISH" BioInteractive, 2013 PDF file.

"International Human Genome Sequencing Consortium Publishes Sequence and Analysis of the Human Genome." National Human Genome Research Institute. 2001 Web. September 2015.

Ji, Q.; Luo, Z-X; Yuan, C-X; and Tabrum, A.R. 2006. 'A Swimming Mammaliaform from the Middle Jurassic and Ecomorphological Diversification of Early Mammals', Science, 311, 1123-7.

Karami, A. "Largest and Smallest Genome in the World" Research Center of Molecular Biology, Baqiyatallah University of Medical Science, 2013 PDF file.

Keller, A. et al. "New insights into the Tyrolean Iceman's origin and phenotype as inferred by whole-genome sequencing." Nature Communications 2012, 3:698 | DOI: 10.1038/ncomms1701. Macmillan Publishers Limited, 2012. PDF file.

Kelly, F. C. 1989. *The Wright Brothers: A Biography*. Mineola: Dover Publications

Lordkipanidze, D.; Ponce de León, M.S.; Margvelashvili, A.; Rak, Y.; Rightmire, G.P.; Vekua, A.; and Zollikofer, C.P.E. 2013. 'A Complete Skull from Dmanisi, Georgia, and the Evolutionary Biology of Early Homo', Science, 342, 326-331.

Lunine, J. I. 2013. *Earth: Evolution of a Habitable World*. Cambridge: Cambridge University Press

Mayor, A. 2011. *The First Fossil Hunters: Dinosaurs, Mammoths, and Myth in Greek and Roman Times*. Princeton: Princeton University Press

Nye, B. 2014. *Undeniable: Evolution and the Science of Creation*. New York: St. Martin's Press

"Ocean." National Oceanic and Atmospheric Administration. Web. September 2015.

Paabo, S. 2014. *Neanderthal Man: In Search of Lost Genomes*. New York: Basic Books

Potten, C. and Wilson, J. 2004. *Apoptosis: The Life and Death of Cells*. Cambridge: Cambridge University Press

Russell, P. J. 2006. *iGenetics: A Molecular Approach*. San Francisco: Pearson/Benjamin Cummings

Scotland, R. and Pennington, R. T. 2000. *Homology and Systematics: Coding Characters for Phylogenetic Analysis*. Boca Raton: CRC Press

"Species." Ensembl. 2015. Web. September 2015

"Species preservation and population size: when eight is not enough." Understanding Evolution. Berkeley, University of California. Web. August 2015

Thomson, K. S. 1992. *Living Fossil: The Story of the Coelacanth*. New York: W. W. Norton & Company

Watts, N. 2000. *The Oxford Greek Dictionary: Greek-English, English-Greek*. New York: Berkley Books

chapter 4.
WHAT EMPIRICAL DATA SUPPORTS EVOLUTION?

Arroyo, E. "Urban Edge Effects and their relationship with the natural environment." California State Parks Online. September 2000. Web. 14 April 2015.

Batten, D. *"Living fossils: a powerful argument for creation."* Creation Magazine. April 2011: 20-23. Print.

Baumann, S. 2002. *Geology Super Review*. Piscataway: Research & Education Assoc.

Bowman, C.; Mathis, A. "The Grand Age of Rocks: The Numeric Ages for Rocks Exposed within Grand Canyon." National Park Service U.S. Department of the Interior. 2006. Web. 14 April 2015

Coyne, J.A. 2015. *Faith Versus Fact: Why Science and Religion Are Incompatible*. New York: Penguin Group.

Coyne, J. A. "Why Evolution is True (But Not Many People Believe It)." Harvard Museum of Natural History. 2 May 2012.

Coyne, J.A. 2009. *Why Evolution is True*. New York: Penguin Group.

Darwin, C. 1872. *The Origin of Species by Means of Natural Selection, or the Preservation of Favoured Races in the Struggle for Life*, 6th edn. London: John Murray

Dawkins, R. 1986. *The Blind Watchmaker: Why the Evidence of Evolution Reveals a Universe Without Design*. New York: W. W. Norton & Company

Dawkins, R. 2009. *The Greatest Show on Earth: the Evidence for Evolution*. New York: Free Press

"Defining Evolution." National Center for Science Education. 2008 Web. September 2015.

Dennett, C. 1995. *Darwin's Dangerous Idea: Evolution and the Meaning of Life*. New York: Touchstone

Deutsch, D. "A new way to explain explanation." TED. Oxford, England. October 2009.

Dickerson, Richard E. 1992 The Game of Science. *Journal of Molecular Evolution* 34: 277-279.

"Discovery of Bacteria That Hasn't Evolved in 2 Billion Years Is New Validation of Darwin's Theory." ABC News. n.d. n.p. Web. 4 Feb. 2015.

"groupthink." wikipedia.org. Wikipedia, 2015. Web. January 2015.

Janis, I. L. 1972. *Victims of Groupthink: A Psychological Study of Foreign-Policy Decisions and Fiascoes*. Boston: Houghton Mifflin Company

Jones, J. E. 2010. *Ruling Against the Dover Area (PA) School District for Requiring That Intelligent Design be Taught as Part of the Biology Curriculum*. Darby: Diane Publishing

Kitzmiller v. Dover Area School District. Trial Transcript: Day 21 am. Session, Part 1 & 2. 4 Nov. 2005.

Knevitt, O. "Let's hope we don't have another Archaeoraptor on our hands." Science 2.0, 2013. Web. November 2015.

"Malaria and the Red Cell." Harvard Information Center for Sickle Cell and Thalassemic Disorders. n.p. 2 April. 2002 Web. 14 April. 2015.

Medema, S. G. 1997. *Coasean Economics Law and Economics and the New Institutional Economics: Law and Economics and the New Institutional Economics.* Berlin: Springer Science & Business Media

"NSTA Position Statement: The Teaching of Evolution." National Science Teachers Association. n.p. July 2013. Web. 14 April. 2015.

"Paleozoic sedimentary rocks." United States Geological Survey. n.p. Web. 14 April. 2015.

Pickrell, J. "How Fake Fossils Pervert Paleontology [Excerpt]." Scientific American. 15 November. 2014. Web. 14 April. 2015

Popper, K. 2005. *The Logic of Scientific Discovery*. London: Routledge

"prediction." Merriam-Webster.com. Merriam-Webster, 2015. Web. January 2015.

Rudwick, M. J. S. 2010. *Worlds Before Adam: The Reconstruction of Geohistory in the Age of Refor*m. Chicago: University of Chicago Press

"science." merriam-webster.com. Webster, 2015. Web. January 2015.

"scientific method." merriam-webster.com. Webster, 2015. Web. January 2015.

"scientific theory." rationalwiki.org. RationalWiki, 2015. Web. January 2015.

Shubin, N. 2008. *Your Inner Fish: A Journey into the 3.5-Billion-Year History of the Human Body*. New York: Knopf Doubleday Publishing Group

Shmueli, G. "To Explain or to Predict?" Statistical Science 2010, Vol. 25, No. 3, 289–310. Institute of Mathematical Statistics, 2010. PDF file.

Surowiecki, J. 2005. *The Wisdom of Crowds*. New York: Knopf Doubleday Publishing Group

Travis, C. B. 2003. *Evolution, Gender, and Rape*. Cambridge: MIT Press

Vandergast, A. G. et al. "Understanding the genetic effects of recent habitat fragmentation in the context of evolutionary history: phylogeography and landscape genetics of a southern California endemic Jerusalem cricket (Orthoptera: Stenopelmatidae: Stenopelmatus)" San Diego State University Online. Blackwell Publishing Ltd. 30 October 2006. Web. 14 April 2015.

Walker, J. "The Biology Primer." Orion Scientific. Web. 14 April 2015.

chapter 5.
IS FAITH NECESSARY?

American Society of Magazine Editors. 2005. *The Best American Magazine Writing*. New York: Columbia University Press

Coyne, J. A. "Why Evolution is True (But Not Many People Believe It)." Harvard Museum of Natural History. 2 May 2012.

Coyne, J.A. 2009. *Why Evolution is True*. New York: Penguin Group.

Dawkins, R. "Militant atheism." TED. Monterey, CA. February 2002.

Dawkins, R. 2009. *The Greatest Show on Earth: the Evidence for Evolution*. New York: Free Press

Dawkins, R. "We would be better off without religion" Methodist Central Hall Westminster. 27 March 2007.

"faith." merriam-webster.com. Webster, 2015. Web. January 2015.

Ham, K. and Nye, B. "Is Creation A Viable Model of Origins?" Answers in Genesis. Legacy Hall, Petersburg, KY. 4 Feb. 2014. Debate.

Kitcher, P. 2007. *Living with DARWIN*. New York: Oxford University Press

Marx, K. 1977. *Critique of Hegel's 'Philosophy Of Right'*. New York: CUP Archive

New International Version. [Colorado Springs]: Biblica, 2011. BibleGateway.com. Web. Jan. 2015.

Newport, F. "In U.S., 42% Believe Creationist View of Human Origins." Gallup Online. 2 June. 2014. Web. 14 April 2015.

"NSTA Position Statement: The Teaching of Evolution." National Science Teachers Association. n.p. July 2013. Web. 14 April. 2015.

"numinous." merriam-webster.com. Webster, 2015. Web. January 2015.

Oaklander, L. N. and Smith, Q. 2005. *Time, Change and Freedom: An Introduction to Metaphysics*. London: Routledge

"religion." merriam-webster.com. Webster, 2015. Web. January 2015.

"Standing Up in the Milky Way." Cosmos: A Spacetime Odyssey. Fox. National Geographic Channel. 9 Mar. 2014. Television.

"supernatural." merriam-webster.com. Webster, 2015. Web. January 2015.

Travis, C. B. 2003. *Evolution, Gender, and Rape*. Cambridge: MIT Press

chapter 6.
IS SPENDING TAX DOLLARS ON EVOLUTION LEGAL?

Coyne, J. A. (2012), SCIENCE, RELIGION, AND SOCIETY: THE PROBLEM OF EVOLUTION IN AMERICA. Evolution, 66: 2654–2663. doi: 10.1111/j.1558-5646.2012.01664.x

Coyne, J. A. "Why Evolution is True (But Not Many People Believe It)." Harvard Museum of Natural History. 2 May 2012.

Dawkins, R. Interview by Paul Hoffman. Big Think. Big Think, 2009. Web. 21 Oct. 2009.

Jefferson, T. 1900. *The Jeffersonian Cyclopedia: A Comprehensive Collection of the Views of Thomas Jefferson Classified and Arranged in Alphabetical Order Under Nine Thousand Titles Relating to Government, Politics, Law, Education, Political Economy, Finance, Science, Art, Literature, Religious Freedom, Morals, Etc.*
New York: Funk & Wagnalls Company

John E. PELOZA v. CAPISTRANO UNIFIED SCHOOL DISTRICT. United States Court of Appeals, Ninth Circuit. 4 Oct. 1994.

Mazur, A. (2004), Believers and disbelievers in evolution. Politics and the Life Sciences 23(2):55-61. doi: 10.2990/1471-5457(2004)23[55:BADIE]2.0.CO;2

National Aeronautics and Space Administration. Fiscal Year 2015 Budget Estimates. NASA 2015. PDF file.

"National Science Foundation fiscal year 2016 budget request continues commitment to discovery, innovation and learning." National Science Foundation. n.p. 2 February 2015. Web. 14 April. 2015.

Newport, F. "More Than 9 in 10 Americans Continue to Believe in God." Gallup Online. 3 June. 2011. Web. 14 April 2015.

"NIH Budget" National Institutes of Health. n.p. 29 January 2015. Web. 14 April. 2015.

Wagman, R. J. 1991. *The First Amendment book*. New York: World Almanac Books

Wells, J. 2002. *Icons of Evolution: Science or Myth? Why Much of What We Teach About Evolution Is Wrong.* Washington DC: Regnery Publishing

Segraves v. California, No. 278978 (Super. Ct. Sacramento County 1981)

CONCLUSION

Coyne, J. A. (2012), SCIENCE, RELIGION, AND SOCIETY: THE PROBLEM OF EVOLUTION IN AMERICA. Evolution, 66: 2654–2663. doi: 10.1111/j.1558-5646.2012.01664.x

Feynman, R. P. & Robbins, J. 2005. *The Pleasure of Finding Things Out: The Best Short Works of Richard P. Feynman.* New York: Basic Books

Jones, J. E. 2010. *Ruling Against the Dover Area (PA) School District for Requiring That Intelligent Design be Taught as Part of the Biology Curriculum.* Darby: Diane Publishing

Kitzmiller v. Dover Area School District. Trial Transcript: Day 21 am. Session, Part 1 & 2. 4 Nov. 2005.

"*Michio Kaku on Singularity 1 on 1: Science is the Engine of Prosperity!.*" Singularity Weblog. 6 June. 2014. Web. 14 April 2015.

Segraves v. California, No. 278978 (Super. Ct. Sacramento County 1981)

"supernatural." merriam-webster.com. Webster, 2015. Web. January 2015.

Thank You.

"To those that have supported this project either publicly or anonymously, I appreciate and thank you from the bottom of my heart."

William James Herath

Made in the USA
Middletown, DE
21 February 2019